FOUNDATIONS OF
ABSTRACT MATHEMATICS

International Series in Pure and Applied Mathematics

Churchill / Brown Series

FOUNDATIONS OF
ABSTRACT MATHEMATICS

David C. Kurtz

Rollins College

McGraw-Hill, Inc.
New York St. Louis San Francisco Auckland Bogotá
Caracas Lisbon London Madrid Mexico Milan
Montreal New Delhi Paris San Juan Singapore
Sydney Tokyo Toronto

This book was set in Times Roman by Publication Services.
The editors were Richard Wallis and Margery Luhrs;
the production supervisor was Denise L. Puryear.
The cover was designed by Karen K. Quigley.
R. R. Donnelley & Sons Company was printer and binder.

FOUNDATIONS OF ABSTRACT MATHEMATICS

2 3 4 5 6 7 8 9 0 DOH DOH 9 0 9 8 7 6 5 4 3 2

ISBN 0-07-035690-4

Library of Congress Cataloging-in-Publication Data

Kurtz, David.
    Foundations of abstract mathematics / by David Kurtz.
       p.   cm. — (International series in pure and applied
mathematics)
    Includes index.
    ISBN 0-07-035690-4
    1. Logic, Symbolic and mathematical.   2. Mathematics.   I. Title.
II. Series.
QA9.K797   1992   91-338
511.3—dc20

ABOUT THE AUTHOR

David C. Kurtz is Professor of Mathematical Sciences and Department Chair at Rollins College. He received a B.S. (from Purdue) and an M.S. (from MIT) in Electrical Engineering and an M.A. (from Wake Forest) and a Ph.D. (from Duke under L. Carlitz) in Mathematics. He has taught undergraduate mathematics for more than 20 years, including 3 years at the University of Malaŵi under a Fulbright Grant. He is the author of several publications in the area of combinatorics and has been active in the calculus renewal movement.

To Judy

CONTENTS

PREFACE

One of the most difficult steps a student of mathematics must make is the one into that (blissful) state known as "mathematical maturity." This is a step which is accomplished by making the transition from solving problems in a fairly concrete setting in which there is a well-known method or an algorithm for each problem type (as in most calculus courses, for example) to writing proofs and producing counterexamples involving more abstract objects and concepts, an activity for which there is no well-defined algorithm. Often this transition is something which is expected just to "happen," perhaps during the summer between the sophomore and junior years; however, it is not clear what summertime activities one could recommend to ensure such a result. My recent teaching experience suggests that this transition is not an easy one for most students and generally cannot be successfully made without some concerted effort and guidance. Two things which seem to inhibit a smooth transition are a lack of knowledge of some fundamental mathematical ideas—logic, sets, functions—and a lack of experience in two important mathematical activities—finding examples of objects with specified properties and writing proofs. This book is an attempt to provide an opportunity to gain exposure to these activities while learning some of the necessary fundamental ideas.

I have tried to keep the book as short as possible to achieve these goals; thus some interesting topics are left out and others are treated only in the exercises. I have also tried to take a developmental point of view so that the book starts out in a fairly simple, informal manner and gradually becomes more formal and abstract. This means that while it is possible to cover the first chapter rather rapidly, one should not expect to maintain this speed throughout the book; indeed, I have found that some sections in Chapter 2 can easily take more than a week to cover with any degree of thoroughness.

The transitional process begins with an informal introduction to logic, including a careful consideration of quantifiers and a discussion of basic proof forms. The objective here is to obtain a firm foundation on which to build the proof-writing skills which will be developed later on. The first mathematical objects encountered in any detail are sets. This provides a familiar setting in which students can write simple proofs, test conjectures and produce counterexamples. The next topic, relations, is probably not as familiar as the topic of sets, and here the students get their first taste of trying to understand the definition of a new concept (e.g., equivalence relation, strict partial order) well enough to provide examples and proofs. Functions are presented as special relations and functional composition is emphasized. Chapter 2 concludes with binary operations and equivalence relations induced by functions, a foreshadowing of the fundamental theorem of group homomorphisms. Several forms of mathematical induction are presented in chapter 3, and since the students should have acquired a working knowledge of implications, propositional functions and sets by this time, there is some chance that their understanding of induction will be more than an algorithmic one.

These first three chapters form what I think should be the core of the course; in fact, with some difficulty I have been restrained from calling it "What Every Mathematics Student Should Know." As time permits (and it sometimes actually does), any of the last three chapters may be studied independently of one another in accordance with the interests and needs of the class. Each is reasonably self-contained and chapters 5 (Groups) and 6 (Cardinality) do not require any previous knowledge and have the advantage of presenting material which is new to the students. Chapter 4 (Continuity Carefully Considered) probably should not be attempted by students who have not had a year of calculus. It begins with a development of the real number system, including algebraic order and metric properties. Extensive use is made of sequences in examining the concepts of limits, continuity and uniform continuity. In chapter 5, cosets are discussed in some detail (as examples of partitions) and their connection to homomorphisms is explored. In chapter 6 much use is made of one-to-one correspondences. The properties of finite and infinite sets are distinguished and cardinal numbers are discussed. In each of these "application" chapters no attempt has been made to give a comprehensive view of the subject being considered; rather, a small area has been examined in sufficient depth so that some non-trivial results can be shown (e.g., intermediate value theorem, fundamental theorem of group homomorphisms, the uncountability of $\mathbb{R}$).

One final comment: In days gone by I thought that if I could organize the material to be presented in a cogent fashion, develop the students' interest in it and provide good examples and answers to their questions, I would be a good teacher and they would learn a lot of mathematics. That is, I thought that what I did was the important part of the educational process. Now I have come to believe that what *I* do is not nearly so important as what I can get the *students* to do. This means that it is impossible to overemphasize the importance of having the students do exercises. I have provided a wide selection of exercises, many of which are presented as conjectures to be verified or shown incorrect. A somewhat unusual sort of exercise which appears throughout the last five chapters of the book is the "Believe It or Not" exercise. In these exercises a conjecture is given, along with a proof and a counterexample. Of course, at least one must be incorrect (sometimes all three are), and the student's task is to sort things out and put them right by determining the true state of affairs and giving (if necessary) a correct proof or counterexample and pointing out the errors in the ones given.

I am not sure if it is possible to *teach* someone how to write proofs, any more than it is possible to teach them how to write poetry or compose a symphony. However, I do think that it is possible to help someone *learn* how to write proofs and I hope that this book is useful in accomplishing this important task.

I am grateful to the many students who, over the years, labored through the large number of iterations of this material. In many cases they have inspired the interesting, but incorrect, parts of the "Believe It or Not" exercises. I want to thank the reviewers whose helpful comments have improved the exposition and helped root out unclear passages:

Barbara Bohannon, Hofstra University;
Richard E. Chandler, North Carolina State University;
Harvey Charlton, North Carolina State University;
Charles Clever, South Dakota State University;
Peter Colwell, Iowa State University;
Gary D. Crown, Wichita State University;
Bruce Edwards, University of Florida;
Robert O. Gamble, Winthrop College;
John I. Gimbel, University of Alaska;
Stephen Pennell, University of Lowell.

Of course any inaccuracies and opaqueness remaining are mine. I am also indebted to the staff at McGraw-Hill who have made the production of this book a pleasure.

David C. Kurtz

A FEW WORDS
FOR THE READER

Many students have difficulty when they are first asked to prove theorems in mathematics. Part of this difficulty may come from an unfamiliarity with the mathematical objects involved (vectors, bases, linear transformations, groups, homomorphisms, and so forth), but a major part of the difficulty seems to be due to an imprecise knowledge of the fundamentals of mathematics: logic, sets, relations and functions. This book attempts to address this problem by giving a concise account of a minimal amount of this material needed to progress further in mathematics and then using this material as a vehicle for gaining practice in proving theorems.

The key word here is *practice*. As you no doubt have observed, learning how to write out a correct proof yourself is quite a bit different from watching someone else write out a proof and understanding that his or her proof is correct. Mathematics is not a spectator sport! Practice and involvement are essential. If anything is to be gained from this book, the reader must become actively engaged in working his or her way through it. This means marking up the pages with questions about unclear passages (should there be any!), doing the examples and then checking the results, working all the exercises and, above all, approaching the subject matter with a questioning mind intent upon gaining a thorough understanding of it.

A passive approach is doomed to failure. A pencil and paper should be at hand before you start reading. Of course, this means that you won't be able to read 20 pages a night; 3 pages would be a more reasonable goal, especially further along in the book where the level of abstraction is somewhat higher and more is expected of you. But as in anything where a considerable effort is required, the rewards are equally great; the satisfaction of writing a proof which you *know* is correct is hard to match. So pick up your pencil (or pen or whatever it is you use) and proceed at a deliberate pace through the following pages, knowing that mastery of their contents will lead to mathematical pleasures unknown to the uninitiated.

FOUNDATIONS OF
ABSTRACT MATHEMATICS

CHAPTER
1

LOGIC

1.1 INTRODUCTION

A friend of mine recently remarked that when he studied logic he got sleepy.
I replied that he looked sleepy at the moment and he said, "Yes, I am sleepy."
He added, "Therefore, you can conclude that I have been studying logic."
"Most certainly not!" I answered. "That's a good example of an invalid
argument. In fact, if you have been studying logic it's obvious that you
haven't learned very much."

This short excerpt from a real-life situation is meant to illustrate the
fact that we use logic in our everyday lives—although we don't always use
it correctly. Logic provides the means by which we reach conclusions and
establish arguments. Logic also provides the rules by which we reason in
mathematics, and to be successful in mathematics we will need to understand
precisely the rules of logic. Of course, we can also apply these rules to areas
of life other than mathematics and amaze (or dismay) our friends with our
logical, well-trained minds.

In this chapter we will describe the various connectives used in logic,
develop some symbolic notation, discover some useful rules of inference,
discuss quantification and display some typical forms of proof. Although
our discussion of connectives and truth tables in the beginning is rather
mechanistic and does not require much thought, by the end of the chapter
we will be analyzing proofs and writing some of our own, a very non-
mechanistic and thoughtful process.

1.2 AND, OR, NOT, AND TRUTH TABLES

The basic building blocks of logic are *propositions*. By a proposition we will mean a declarative sentence which is either true or false but not both. For example, "2 is greater than 3" and "All equilateral triangles are equiangular" are propositions while "$x < 3$" and "This sentence is false" are not (the first of these is a declarative sentence but we cannot assign a truth value until we know what "x" represents; try assigning a truth value to the second). We will denote propositions by lowercase letters, p, q, r, s, etc. In any given discussion different letters may or may not represent different propositions but a letter appearing more than once in a given discussion will always represent the same proposition. A true proposition will be given a truth value of T (for true) and a false proposition a truth value of F (for false). Thus "$2 + 3 < 7$" has a truth value of T while "$2 + 3 = 7$" has a truth value of F.

We are interested in combining simple propositions (sometimes called *subpropositions*) to make more complicated (or compound) propositions. We combine propositions with *connectives,* among which are "and," "or" and "implies." If p, q are two propositions then "p and q" is also a proposition, called the *conjunction* of p and q, and denoted by

$$p \wedge q.$$

The truth value of $p \wedge q$ depends on the truth values of the propositions p and q: $p \wedge q$ is true when p and q are both true, otherwise it is false. Notice that this is the usual meaning we assign to "and." The word "but" has the same logical meaning as "and" even though in ordinary English it carries a slightly different connotation. A convenient way to display this fact is by a *truth table*. As each of the two propositions p, q has two possible truth values, together they have $2 \times 2 = 4$ possible truth values so the table below lists all possibilities:

p	q	$p \wedge q$
T	T	T
T	F	F
F	T	F
F	F	F

Thus, for example, when p is T and q is F (line 2 of the truth table), $p \wedge q$ is F. In fact, this truth table can be taken as the definition of the connective $\wedge$.

It should be noted here that the truth table above does not have anything to do with p and q; they are just placeholders—cast in the same role as x in the familiar functional notation $f(x) = 2x - 3$. What the truth table does tell us, for example, is that when the first proposition is F and the second is T (third row of the table) the conjunction of the two propositions is F. You can check your understanding of this point by working exercise 5 at the end of this section.

Another common connective is "or," sometimes called *disjunction*. The disjunction of p and q, denoted by

$$p \lor q$$

is true when *at least one* of p, q is true. This is called the "inclusive or"; it corresponds to the "and/or" sometimes found in legal documents. Note that in ordinary conversation we often use "or" in the exclusive sense; true only when *exactly one* of the subpropositions is true. For example, the truth of "When you telephoned I must have been in the shower or walking the dog" isn't usually meant to include both possibilities. In mathematics we always use "or" in the inclusive sense as defined above and given in the truth table below:

p	q	$p \lor q$
T	T	T
T	F	T
F	T	T
F	F	F

Given any proposition p we can form a new proposition with the opposite truth value, called the *negation* of p, which is denoted by

$$\neg p.$$

This is sometimes read as "not p."

The truth table for negation is

p	$\neg p$
T	F
F	T

We can form the negation of a proposition without understanding the meaning of the proposition by prefacing it with "it is false that" or "it is not the case that" but the resulting propositions are usually awkward and do not convey the real nature of the negation. A closer consideration of the meaning of the proposition in question will often indicate a better way of expressing the negation; later we will consider methods for negating compound propositions.

Consider the examples below:

a) $3 + 5 > 7$.
b) It is not the case that $3 + 5 > 7$.
c) $3 + 5 \leq 7$.
d) $x^2 - 3x + 2 = 0$ is not a quadratic equation.
e) It is not true that $x^2 - 3x + 2 = 0$ is not a quadratic equation.
f) $x^2 - 3x + 2 = 0$ is a quadratic equation.

Note that b) and c) are negations of a); e) and f) are negations of d), but c) and f) are to be preferred over b) and e), respectively.

We will use the same convention for $\neg$ as we use for $-$ in algebra; that is, it applies only to the next symbol, which in our case represents a proposition. Thus $\neg p \vee q$ will mean $(\neg p) \vee q$ rather than $\neg(p \vee q)$, just as $-3 + 4$ represents 1 and not -7. With this convention we can be unambiguous when we negate compound propositions using symbols, but life is not so easy when we consider how to negate compound propositions in English. For example, how do we distinguish between $\neg p \vee q$ and $\neg(p \vee q)$ in English? Suppose p represents "$2 + 2 = 4$," and q represents "$3 + 2 < 4$." Should "It is not the case that $2 + 2 = 4$ or $3 + 2 < 4$" mean $\neg(p \vee q)$ or $\neg p \vee q$? If we use the same convention we used for our symbols it should mean $\neg p \vee q$. But, if we take this meaning, then how would we say $\neg(p \vee q)$? The problem seems to be a lack of the equivalent of the parentheses we used for grouping. Let us adopt the convention that "it is not the case that" (or a similar negating phrase) applies to everything that follows, up to some sort of grouping punctuation. Thus, "It is not the case that $2 + 2 = 4$ or $3 + 2 < 4$" would mean $\neg(p \vee q)$, while "It is not the case that $2 + 2 = 4$, or $3 + 2 < 4$" would mean $\neg p \vee q$. Of course, when speaking, one must be very careful about using pauses to indicate the proper meaning.

Truth tables can be used to express the possible truth values of compound propositions by constructing the various columns in a methodical manner. For example, suppose that we wish to construct the truth table for $\neg(p \vee \neg q)$. We begin by making a basic four-row (there are four possibilities) truth table with column headings:

p	q	$\neg$	(	p	$\vee$	$\neg$	q	)
T	T							
T	F							
F	T							
F	F							

Truth values are then entered step by step:

p	q	$\neg$	(	p	$\vee$	$\neg$	q	)
T	T			T			T	
T	F			T			F	
F	T			F			T	
F	F			F			F	

p, q columns entered

p	q	$\neg$	(	p	$\vee$	$\neg$	q	)
T	T			T		F	T	
T	F			T		T	F	
F	T			F		F	T	
F	F			F		T	F	

$\neg q$ column entered

p	q	$\neg$	(	p	$\vee$	$\neg$	q	)
T	T			T	T	F	T	
T	F			T	T	T	F	
F	T			F	F	F	T	
F	F			F	T	T	F	

$p \vee \neg q$ column entered

p	q	$\neg$	(	p	$\vee$	$\neg$	q	)
T	T	**F**		T	T	F	T	
T	F	**F**		T	T	T	F	
F	T	**T**		F	F	F	T	
F	F	**F**		F	T	T	F	

$\neg(p \vee \neg q)$ column entered

After some experience is obtained, many of the steps written above can be eliminated. We also note that if a compound proposition involves n subpropositions then its truth table will require 2^n rows. Thus a compound proposition with four subpropositions will require $2^4 = 16$ rows.

Exercises 1.2

1. Assign truth values to the following propositions:
   a) $3 \le 7$ and 4 is an odd integer.
   b) $3 \le 7$ or 4 is an odd integer.
   c) $2 + 1 = 3$ but $4 < 4$.
   d) 5 is odd or divisible by 4.
   e) It is not true that $2 + 2 = 5$ and $5 > 7$.
   f) It is not true that $2 + 2 = 5$ or $5 > 7$.
   g) $3 \ge 3$.

2. Suppose that we represent "7 is an even integer" by p, "$3 + 1 = 4$" by q and "24 is divisible by 8" by r.
   a) Write the following in symbolic form and assign truth values:
      i) $3 + 1 \ne 4$ and 24 is divisible by 8.
      ii) It is not true that 7 is odd or $3 + 1 = 4$.
      iii) $3 + 1 = 4$ but 24 is not divisible by 8.
   b) Write out the following in words and assign truth values:
      i) $p \vee \neg q$.
      ii) $\neg(r \wedge q)$.
      iii) $\neg r \vee \neg q$.

3. Construct truth tables for
   a) $\neg p \vee q$.
   b) $\neg p \wedge p$.
   c) $(\neg p \vee q) \wedge r$.
   d) $\neg(p \wedge q)$.
   e) $\neg p \wedge \neg q$.
   f) $\neg p \vee \neg q$.

g) $p \vee \neg p$.

h) $\neg(\neg p)$.

4. Give useful negations of

a) $3 - 4 < 7$.

b) $3 + 1 = 5$ and $2 \le 4$.

c) 8 is divisible by 3 but 4 is not.

5. Suppose that we define the connective $\star$ by saying that $p \star q$ is true only when q is true and p is false and is false otherwise.

a) Write out the truth table for $p \star q$.

b) Write out the truth table for $q \star p$.

c) Write out the truth table for $(p \star p) \star q$.

6. Let us denote the "exclusive or" sometimes used in ordinary conversation by $\oplus$. Thus $p \oplus q$ will be true when exactly one of p, q is true, and false otherwise.

a) Write out the truth table for $p \oplus q$.

b) Write out the truth tables for $p \oplus p$ and $(p \oplus q) \oplus q$.

c) Show that "and/or" really means "and or or," that is, the truth table for $(p \wedge q) \oplus (p \oplus q)$ is the same as the truth table for $p \vee q$.

d) Show that it makes no difference if we take both "or's" in "and/or" to be inclusive ($\vee$) or exclusive ($\oplus$).

1.3 IMPLICATION AND THE BICONDITIONAL

If we were to write out the truth tables for $\neg(p \wedge q)$ and for $\neg p \vee \neg q$ (as we did in exercise 3 d), f) above) and compare them, we would note that these two propositions have the same truth values and thus in some sense are the same. This is an important concept (important enough to have a name anyway) so we make the following definition:

Suppose that two propositions p, q have the same truth table. Then p and q are said to be *logically equivalent,* which we will denote by

$$p \Longleftrightarrow q.$$

Basically, when two propositions are logically equivalent they have the same form and we may substitute one for the other in any other proposition or theorem. It is important to emphasize that it is the *form* and not the truth value of a proposition which determines whether it is (or is not) logically equivalent to another proposition. For example, "$2 + 2 = 4$" and "$7 - 5 = 2$" are both true propositions but they are *not* logically equivalent since they have different truth tables (if we represented the first by p then the other would need another symbol, say q, and we know that these do not have the same truth tables). On the other hand, "2 + 3 = 5 or 3 − 4 = 2" and "3 − 4 = 2 or 2 + 3 = 5" *are* logically equivalent. To see this, let p represent "3 − 4 = 2" and q represent "2 + 3 = 5." Then the first is of the form $q \lor p$ while the second has the form $p \lor q$. A check of truth tables shows that these two do indeed have the same truth table.

Using this idea of logical equivalence we can state the relationship between negation, disjunction and conjunction, sometimes called *DeMorgan's laws*:

Let p, q be any propositions. Then

$$\neg(p \lor q) \Longleftrightarrow \neg p \land \neg q,$$

$$\neg(p \land q) \Longleftrightarrow \neg p \lor \neg q.$$

We have already verified the second of these in exercise 3 d) and f) in the previous section; the reader should verify the other by means of a truth table now. In words, DeMorgan's laws state that the negation of a disjunction is logically equivalent to the conjunction of the negations and the negation of a conjunction is logically equivalent to the disjunction of the negations. A common mistake is to treat $\neg$ in logic as $-$ in algebra and to think that $\neg$ distributes over $\lor$ and $\land$ just as $-$ distributes over $+$. That is, since $-(a + b) = -a + (-b)$, one might be led to believe, for example, that $\neg(p \lor q) \Leftrightarrow \neg p \lor \neg q$. A quick check with truth tables (or reference to exercise 3 d), e) in the previous section) shows that this isn't correct. Thus, while our logical notation appears somewhat "algebra-like" (and indeed is an example of a certain kind of algebra) its rules differ from those of the familiar algebra of real numbers and we should not make the mistake of assuming that certain logical operations behave in ways analogous to our old algebraic friends $+$, $\times$ and $-$.

One of the most important propositional forms in mathematics is that of implication, sometimes called the *conditional*. In fact, all mathematical theorems are in the form of an implication: If "hypothesis" then "conclusion." The general form of an implication is "if p then q" where p, q are propositions; we will denote this by

$$p \rightarrow q.$$

In the conditional $p \rightarrow q$, p is called the *premise* (or hypothesis or antecedent) and q is called the *conclusion* (or *consequence* or *consequent*). The truth table for $p \rightarrow q$ is

p	q	$p \rightarrow q$
T	T	T
T	F	F
F	T	T
F	F	T

If we think about the usual meaning we give to *implies* we should agree that the first two lines of the above truth table correspond to ordinary usage, but that the last two lines may not be so clear. Of course, we are free to define the truth values of the various connectives in any way we choose and we could take the position that this is the way we want to define *implies* (which is indeed the case) but it is worthwhile to see that the definition above also agrees with everyday usage. To this end, let us consider what might be called "The parable of the dissatisfied customer." Imagine that we have purchased a product, say a washday detergent called *Tyde*, after hearing an advertisement which said, "If you use Tyde then your wash will be white!" Under what circumstances could we complain to the manufacturer? A little thought reveals that we certainly couldn't complain if we had not used Tyde (the ad said nothing about what would happen if we used Chear, for example), and we couldn't complain if we used Tyde and our wash was white; thus we could complain only in the case when we had used Tyde and our wash was not white (as promised). Thus, the ad's promise is false only when "we use Tyde and get a non-white wash" is true. Let's use our logic notation to examine this situation more closely. Let p represent "We use Tyde," and q represent "Our wash is white." Then the advertisement's promise is

$$p \rightarrow q$$

and we can complain (that is, this promise is false) only in the case when

$$p \wedge \neg q$$

is true. Thus, $p \wedge \neg q$ should be logically equivalent to $\neg(p \rightarrow q)$. Writing out the truth table for $p \wedge \neg q$ we get (the reader should verify this):

p	q	$p \wedge \neg q$
T	T	F
T	F	T
F	T	F
F	F	F

As this is to be logically equivalent to the negation of $p \to q$, the truth table for $p \to q$ should be the negation of this (which it is—look back to check this) and our logic definition of implication does agree with our everyday (or at least washday) usage.

We note that the only case in which $p \to q$ is false is when p is true and q is false; that is, when the hypothesis is true and the conclusion is false. Thus the following implications are all true:

a) If $2 + 2 = 4$ then $1 + 1 = 2$.

b) If $2 + 3 = 4$ then $1 + 1 = 5$.

c) If green is red then the moon is made of cheese.

d) If green is red then the moon is not made of cheese.

e) $7 < 2$ if $2 < 1$.

It should also be noted that if an implication is true then its conclusion may be true or false (see examples a), b) above), but if an implication is true *and* the hypothesis is true then the conclusion *must* be true. This, of course, is the basic form of a mathematical theorem: if we know the theorem (an implication) is correct (true) and the hypothesis of the theorem is true we can take the conclusion of the theorem to be true.

There are many ways of stating the conditional in English and all the following are considered logically equivalent:

a) If p then q.

b) p implies q.

c) p is stronger than q.

d) q is weaker than p.

e) p only if q.

f) q if p.

g) p is sufficient for q.

h) q is necessary for p.

i) A necessary condition for p is q.

j) A sufficient condition for q is p.

Most of the time we will use the first two, but it is important to be familiar with the rest. Keeping in mind the definition of $p \to q$ will help us to remember some of these. For example, when we say "r is sufficient for s," we mean that the truth of r is sufficient to guarantee the truth of s; that is, we mean $r \to s$. Similarly, if we say "r is necessary for s," we mean that when s is true, r must necessarily be true too; that is, we mean $s \to r$.

When we observe the truth table for $p \to q$ we note that it is not symmetric with respect to p and q; that is, the truth table for $p \to q$ is not the same as the truth table for $q \to p$. In other words, these two propositions are *not* logically equivalent and thus cannot be substituted one for another. Because of this lack of symmetry it is convenient to make the following definitions:

Given an implication $p \to q$:

$q \to p$ is called its *converse,*

$\neg q \to \neg p$ is called its *contrapositive,*

$\neg p \to \neg q$ is called its *inverse.*

Even though the reader has probably already noticed it, it is worth pointing out that the inverse of an implication is the contrapositive of its converse (it is also the converse of its contrapositive).

Perhaps the most common logical error is that of confusing an implication with its converse (or inverse). In fact, this common error seems to be the basis for much advertising. For example, if we are told that "If we use Tyde then our wash will be white!" (which may be true) we are apparently expected to also believe that if we don't use Tyde then our wash won't be white. But this is the inverse, which is logically equivalent to the converse, of the original claim. Thus, we see that we can believe Tyde's claim and still use Chear with a clear (logically, anyway) conscience and wear white clothes. However, an implication and its contrapositive *are* logically equivalent (see exercises below) and thus may be used interchangeably. In this case, this means that if our clothes are not white then we didn't use Tyde.

The final connective which we will consider is the *biconditional*. If p, q are two propositions then the proposition "p if and only if q" (sometimes abbreviated "p iff q"), denoted by

$$p \leftrightarrow q,$$

is called the *biconditional* (not to be confused with logical equivalence "$\iff$," although there is a connection which will be revealed in the next section; keep reading). We say that $p \leftrightarrow q$ is true when p, q have the same truth value and false when they have different truth values. Thus the truth table for the biconditional is

p	q	$p \leftrightarrow q$
T	T	T
T	F	F
F	T	F
F	F	T

Some other ways of expressing $p \leftrightarrow q$ are

> p is necessary and sufficient for q.
> p is equivalent to q.

As the names (biconditional, if and only if) and notation suggest, there is a close connection between the conditional and the biconditional. In fact, $p \leftrightarrow q$ is logically equivalent to $(p \to q) \wedge (q \to p)$.

Exercises 1.3

1. Which of the following are logically equivalent?
 a) $p \wedge \neg q$.
 b) $p \to q$.
 c) $\neg(\neg p \vee q)$.
 d) $q \to \neg p$.
 e) $\neg p \vee q$.
 f) $\neg(p \to q)$.
 g) $p \to \neg q$.
 h) $\neg p \to \neg q$.

2. Show that the following pairs are logically equivalent:
 a) $p \wedge (q \vee r)$; $(p \wedge q) \vee (p \wedge r)$.
 b) $p \vee (q \wedge r)$; $(p \vee q) \wedge (p \vee r)$.
 c) $p \leftrightarrow q$; $(p \to q) \wedge (q \to p)$.
 d) $p \to q$; $\neg q \to \neg p$.

3. Show that the following pairs are not logically equivalent:

a) $\neg(p \wedge q)$; $\neg p \wedge \neg q$.

b) $\neg(p \vee q)$; $\neg p \vee \neg q$.

c) $p \rightarrow q$; $q \rightarrow p$.

d) $\neg(p \rightarrow q)$; $\neg p \rightarrow \neg q$.

4. Find:

a) The contrapositive of $\neg p \rightarrow q$.

b) The converse of $\neg q \rightarrow p$.

c) The inverse of the converse of $q \rightarrow \neg p$.

d) The negation of $p \rightarrow \neg q$.

e) The converse of $\neg p \wedge q$.

5. Indicate which of the following is true:

a) If $2 + 1 = 4$ then $3 + 2 = 5$.

b) Red is white if and only if green is blue.

c) $2 + 1 = 3$ and $3 + 1 = 5$ implies 4 is odd.

d) If 4 is odd then 5 is odd.

e) If 4 is odd then 5 is even.

f) If 5 is odd then 4 is odd.

6. Give examples of or tell why no such example exists:

a) A true implication with a false conclusion.

b) A true implication with a true conclusion.

c) A false implication with a true conclusion.

d) A false implication with a false conclusion.

e) A false implication with a false hypothesis.

f) A false implication with a true hypothesis.

g) A true implication with a true hypothesis.

h) A true implication with a false hypothesis.

7. Translate into symbols:

a) p whenever q.

b) p unless q.

8. Give a negation for $p \leftrightarrow q$ in a form which does not involve a biconditional.

9. Suppose that p, $\neg q$ and r are true. Which of the following is true?

a) $p \rightarrow q$.

b) $q \rightarrow p$.

c) $p \rightarrow (q \vee r)$.

d) $p \leftrightarrow q$.

e) $p \leftrightarrow r$.

f) $(p \vee q) \rightarrow p$.

g) $(p \wedge q) \rightarrow q$.

10. We note that we now have five logic "connectives": $\wedge$, $\vee$, $\rightarrow$, $\leftrightarrow$ and $\neg$, each of which corresponds to a construct from our ordinary language. It turns out that from a logical point of view this is somewhat wasteful, since we could express all these in terms of just $\neg$ and $\wedge$. Even more, if we define $p \mid q$ to be false when both p and q are true and true otherwise, we could express all five forms in terms of this one connective ($\mid$ is known as the Sheffer stroke). Partially verify the statements given above by

   a) Finding a proposition which is equivalent to $p \vee q$ using just $\wedge$ and $\neg$.

   b) Writing out the truth table for $p \mid q$.

   c) Showing that $p \mid p$ is equivalent to $\neg p$.

   d) Showing that $(p \mid q) \mid (q \mid p)$ is equivalent to $p \wedge q$.

1.4 TAUTOLOGIES

An important class of propositions are those whose truth tables contain only T's in the final column; that is, propositions which are always true and the fact that they are always true depends only on their form and not on any meaning which might be assigned to them (for example, recall exercise 3 g) of section 1.2: $p \vee \neg p$). Such propositions are called *tautologies*. It is important to distinguish between true propositions and tautologies. For example, "$2 + 2 = 4$" is a true proposition but it is not a tautology because its form is p which is not always true. On the other hand, "5 is a primitive root of 17 or 5 is not a primitive root of 17" is a tautology no matter what being a primitive root means. It is a tautology by virtue of its form ($p \vee \neg p$) alone.

    The negation of a tautology, that is, a proposition which is always false, is called a *contradiction*. We must distinguish between contradictions and false statements in the same way we distinguish between true statements and tautologies; a proposition is a contradiction based on its form alone. As examples, consider the truth tables:

p	q	p	$\rightarrow$	$(p$	$\vee$	$q)$
T	T	T	T	T	T	T
T	F	T	T	T	T	F
F	T	F	T	F	T	T
F	F	F	T	F	F	F

p	q	$(p$	$\rightarrow$	$q)$	$\wedge$	$(p$	$\wedge$	$\neg q)$
T	T	T	T	T	F	T	F	F
T	F	T	F	F	F	T	T	T
F	T	F	T	T	F	F	F	F
F	F	F	T	F	F	F	F	T

We see that $p \rightarrow (p \vee q)$ is a tautology and $(p \rightarrow q) \wedge (p \wedge \neg q)$ is a contradiction.

Using the idea of tautology, perhaps we can make clear the distinction between "equivalent" and "logically equivalent." Two propositions p, q are logically equivalent if and only if $p \leftrightarrow q$ is a tautology. Actually, $p \leftrightarrow q$ and $p \Longleftrightarrow q$ are propositions on two different levels. If we think of "p is equivalent to q" as a proposition, then "p is logically equivalent to q" is a proposition about this proposition; namely, the (meta)-proposition "p is equivalent to q is true." For example, $(p \rightarrow q) \leftrightarrow (\neg q \rightarrow \neg p)$ is a logical implication while $p \rightarrow (p \wedge q)$ is not; it is "just" an implication which may or may not be true.

We also use the idea of tautology to make the following definition: we say that $p \rightarrow q$ is a *logical implication* (also "p logically implies q" or "q is a logical consequence of p") if $p \rightarrow q$ is a tautology. p logically implies q is denoted by

$$p \Rightarrow q.$$

Note that logical implication bears the same relation to implication as logical equivalence bears to equivalence. If p logically implies q, and p is true, then q must also be true. For example, $p \rightarrow (p \vee q)$, $(p \wedge q) \rightarrow p$ are logical implications while $p \rightarrow (p \wedge q)$ is not (when p is T and q is F this last implication is F and hence not a tautology).

Tautologies form the rules by which we reason and for future reference a list of the more common ones, along with some of their names, is given below (p, q, r represent any propositions, **c** represents any contradiction, **t** represents any tautology).

A list of tautologies

1. $p \vee \neg p$
2. $\neg (p \wedge \neg p)$
3. $p \rightarrow p$
4. a) $p \leftrightarrow (p \vee p)$ idempotent laws
   b) $p \leftrightarrow (p \wedge p)$
5. $\neg \neg p \leftrightarrow p$ double negation
6. a) $(p \vee q) \leftrightarrow (q \vee p)$ commutative laws
   b) $(p \wedge q) \leftrightarrow (q \wedge p)$
   c) $(p \leftrightarrow q) \leftrightarrow (q \leftrightarrow p)$
7. a) $(p \vee (q \vee r)) \leftrightarrow ((p \vee q) \vee r)$ associative laws
   b) $(p \wedge (q \wedge r)) \leftrightarrow ((p \wedge q) \wedge r)$
8. a) $(p \wedge (q \vee r)) \leftrightarrow ((p \wedge q) \vee (p \wedge r))$ distributive laws
   b) $(p \vee (q \wedge r)) \leftrightarrow ((p \vee q) \wedge (p \vee r))$
9. a) $(p \vee c) \leftrightarrow p$ identity laws
   b) $(p \wedge c) \leftrightarrow \mathbf{c}$
   c) $(p \vee \mathbf{t}) \leftrightarrow \mathbf{t}$
   d) $(p \wedge \mathbf{t}) \leftrightarrow p$
10. a) $\neg (p \wedge q) \leftrightarrow (\neg p \vee \neg q)$ DeMorgan's laws
    b) $\neg (p \vee q) \leftrightarrow (\neg p \wedge \neg q)$
11. a) $(p \leftrightarrow q) \leftrightarrow ((p \rightarrow q) \wedge (q \rightarrow p))$ equivalance
    b) $(p \leftrightarrow q) \leftrightarrow ((p \wedge q) \vee (\neg p \wedge \neg q))$
    c) $(p \leftrightarrow q) \leftrightarrow (\neg p \leftrightarrow \neg q)$
12. a) $(p \rightarrow q) \leftrightarrow (\neg p \vee q)$ implication
    b) $\neg (p \rightarrow q) \leftrightarrow (p \wedge \neg q)$
13. $(p \rightarrow q) \leftrightarrow (\neg q \rightarrow \neg p)$ contrapositive
14. $(p \rightarrow q) \leftrightarrow ((p \wedge \neg q) \rightarrow \mathbf{c})$ reductio ad absurdum
15. a) $((p \rightarrow r) \wedge (q \rightarrow r)) \leftrightarrow ((p \vee q) \rightarrow r)$
    b) $((p \rightarrow q) \wedge (p \rightarrow r)) \leftrightarrow (p \rightarrow (q \wedge r))$
16. $((p \wedge q) \rightarrow r) \leftrightarrow (p \rightarrow (q \rightarrow r))$ exportation law
17. $p \rightarrow (p \vee q)$ addition
18. $(p \wedge q) \rightarrow p$ simplification
19. $(p \wedge (p \rightarrow q)) \rightarrow q$ modus ponens
20. $((p \rightarrow q) \wedge \neg q) \rightarrow \neg p$ modus tollens
21. $((p \rightarrow q) \wedge (q \rightarrow r)) \rightarrow (p \rightarrow r)$ hypothetical syllogism
22. $((p \vee q) \wedge \neg p) \rightarrow q$ disjunctive syllogism
23. $(p \rightarrow \mathbf{c}) \rightarrow \neg p$ absurdity
24. $((p \rightarrow q) \wedge (r \rightarrow s)) \rightarrow ((p \vee r) \rightarrow (q \vee s))$
25. $(p \rightarrow q) \rightarrow ((p \vee r) \rightarrow (q \vee r))$

Observe that in the above list, 4-16 are logical equivalences while 17-25 are logical implications.

One of the first questions from students when they see the above list is, "Do we have to memorize all of these?" The answer is, "No, memorization is not sufficient, you need to *know* all these! They need to be incorporated into your way of thinking." At first glance, this may seem like a formidable task, and perhaps it is. But some of these are already incorporated into our way of thinking. For example, if someone says, "This sweater is orlon or wool. It isn't orlon," what do we conclude about the sweater? We conclude that it is a wool sweater, and in doing so we have just used the disjunctive syllogism (22 on the list above). Similarly, if someone says, "If I do the assignments then I enjoy the class. I did the assignment for today," we conclude that the person speaking enjoyed the class today. This is an application of the modus ponens (19 on the list). It is not important that we learn the names of the various equivalences and implications, but it is important that we learn their *forms* so that we can recognize when we are using them. It is also important to recognize when we are not reasoning correctly; that is, when we use something which is not a logical implication. In the next section we will spend some time looking at this point.

Exercises 1.4

1. Verify that 7 a), 9 b), 13 and 14 in the list above are tautologies.

2. Determine which of the following have the form of something on the above list (for example, $(\neg q \wedge p) \to \neg q$ has the form of 18) and in these cases, indicate which one:
   a) $\neg q \to (\neg q \vee \neg p)$.
   b) $q \to (q \wedge \neg p)$.
   c) $(r \to \neg p) \leftrightarrow (\neg r \vee \neg p)$.
   d) $(p \to \neg q) \leftrightarrow \neg(\neg p \to q)$.
   e) $(\neg r \to q) \leftrightarrow (\neg q \to r)$.
   f) $(p \to (\neg r \vee q)) \leftrightarrow ((r \wedge \neg q) \to \neg p)$.
   g) $r \to \neg(q \wedge \neg r)$.
   h) $((\neg q \vee p) \wedge q) \to p$.

3. Give examples of or tell why no such example exists:
   a) A logical implication with a false conclusion.
   b) A logical implication with a true conclusion.
   c) A logical implication with a true hypothesis and false conclusion.

4. Which of the following are correct?
   a) $(p \to (q \vee r)) \Rightarrow (p \to q)$.
   b) $((p \vee q) \to r) \Rightarrow (p \to r)$.
   c) $(p \vee (p \wedge q)) \Leftrightarrow p$.
   d) $((p \to q) \wedge \neg p) \Rightarrow \neg q$.

5. Which of the following are tautologies, contradictions or neither?
   a) $(p \wedge \neg q) \rightarrow (q \vee \neg p)$. ne
   b) $\neg p \rightarrow p$. ne
   c) $\neg p \leftrightarrow p$. contr
   d) $(p \wedge \neg p) \rightarrow p$. t
   e) $(p \wedge \neg p) \rightarrow q$. t
   f) $(p \wedge \neg q) \leftrightarrow (p \rightarrow q)$. contrad
   g) $[(p \rightarrow q) \leftrightarrow r] \leftrightarrow [p \rightarrow (q \leftrightarrow r)]$. neith

6. Which of the following are correct?
   a) $(p \leftrightarrow q) \Rightarrow (p \rightarrow q)$.
   b) $(p \rightarrow q) \Rightarrow (p \leftrightarrow q)$.
   c) $(p \rightarrow q) \Rightarrow q$.

7. Is $\rightarrow$ associative; i.e., is $((p \rightarrow q) \rightarrow r) \Longleftrightarrow (p \rightarrow (q \rightarrow r))$?

8. Is $\leftrightarrow$ associative; i.e., is $((p \leftrightarrow q) \leftrightarrow r) \Longleftrightarrow (p \leftrightarrow (q \leftrightarrow r))$?

9. Which of the following true propositions are tautologies?
   a) If $2 + 2 = 4$ then 5 is odd. yes
   b) $3 + 1 = 4$ and $5 + 3 = 8$ implies $3 + 1 = 4$.
   c) $3 + 1 = 4$ and $5 + 3 = 8$ implies $3 + 2 = 5$.
   d) Red is yellow or red is not yellow.
   e) Red is yellow or red is red.
   f) 4 is odd or 2 is even and 2 is odd implies 4 is odd.
   g) 4 is odd or 2 is even and 2 is odd implies 4 is even.

10. Which of the following are logical consequences of the set of propositions $p \vee q$, $r \rightarrow \neg q$, $\neg p$?
   a) q.
   b) r.
   c) $\neg p \vee s$.
   d) $\neg r$.
   e) $\neg(\neg q \wedge r)$.
   f) $q \rightarrow r$.

1.5 ARGUMENTS AND THE PRINCIPLE OF DEMONSTRATION

How do you win an argument? Aside from intimidation, force of personality, coercion or threats, of course; we are speaking of convincing someone of the logical correctness of your position. You might begin by saying, "Do you accept p, q and r as being true?" If the answer is, "Yes, of course. Any dolt can see that!" then you say, "Well then, it follows that t must be true." For you to win your argument it must be the case (and this is what you must argue) that $(p \wedge q \wedge r) \rightarrow t$ is a tautology; that is, there is no way for your premises (p, q, r, which your friend has also accepted) to all be true *and* your conclusion, t, be false. It is much the same for the proof of a mathematical theorem; in the proof we must show that whenever the premises of the theorem are true, the conclusion is also true. We will try to put this idea on a more formal basis and then discuss some techniques for demonstrating that an argument is correct. As usual, we start with some definitions.

By an *argument* (or theorem) we mean a proposition of the form

$$(p_1 \wedge p_2 \wedge \cdots \wedge p_n) \to q.$$

We will call p_1, p_2, ..., p_n the *premises* (or hypotheses) and q the *conclusion*. An argument is *valid* (or the theorem is true) if it is a tautology. In this case we say that q (the conclusion) is a *logical consequence* of p_1, p_2, ..., p_n (the premises).

Observe that a valid argument is a logical implication. Thinking about the truth table for the implication we see that this means that whenever p_1, p_2, ..., p_n are all true then q must also be true. Viewed in this light, the above definition of valid argument is seen to agree with the meaning we ordinarily use; if the premises are all true *and* the argument is valid then the conclusion *must* be true. Note that if an argument is valid, the conclusion may be true or false; all that is asserted is that *if* the premises are all true *then* the conclusion must be true. For example, consider the following argument:

$$(\neg q \wedge (p \to q)) \to \neg p.$$

A common way to display arguments is to list the premises, draw a horizontal line and then write the conclusion. Thus the argument above would be displayed:

$$\neg q$$
$$\underline{p \to q}$$
$$\neg p.$$

To test the validity of this argument we can use a truth table:

p	q	$(\neg q$	$\wedge$	$p \to q)$	$\to$	$\neg p$
T	T	F	F	T	**T**	F
T	F	T	F	F	**T**	F
F	T	F	F	T	**T**	T
F	F	T	T	T	**T**	T

As the argument is a tautology, it is a valid argument. We note that this means that whenever the premises are all true (in this case line 4), the conclusion is also true.

Now consider this argument:

$$\neg p$$
$$\frac{p \rightarrow q}{\neg q.}$$

Again, we write the truth table:

p	q	$(\neg p$	$\wedge$	$p \rightarrow q)$	$\rightarrow$	$\neg q$
T	T	F	F	T	**T**	F
T	F	F	F	F	**T**	T
F	T	T	T	T	**F**	F
F	F	T	T	T	**T**	T

As this argument is not a tautology (in line 3 we see that the premises are true but the conclusion is false) it is not valid.

To make these examples a little more concrete, let p represent "$2 + 2 = 4$," and let q represent "$3 + 5 = 7$." Then the first argument becomes

$$3 + 5 \neq 7$$
$$\frac{\text{If } 2 + 2 = 4 \text{ then } 3 + 5 = 7}{2 + 2 \neq 4.}$$

The second is

$$2 + 2 \neq 4$$
$$\frac{\text{If } 2 + 2 = 4 \text{ then } 3 + 5 = 7}{3 + 5 \neq 7.}$$

In the first case (a valid argument) we see that the conclusion is false, while in the second case (an invalid argument) the conclusion is true! What is going on here? The answer is that the validity (or lack thereof) of an argument is based only on the *form of the argument* and has nothing to do with the truth or falsity of the propositions involved (if this were not the case, we would not be able to represent it in symbolic form). Also, it is important to remember that the validity of an argument guarantees the truth of the conclusion only when all the premises are all true. In the first argument above we see that the second premise, "If $2 + 2 = 4$ then $3 + 5 = 7$," is false.

Although the above procedure of using truth tables to check the validity of an argument is simple to use and doesn't require much thought, it is not very convenient when the number of propositions is large; for example, if there were eight propositions the truth table would require $2^8 = 256$ rows.

Another method of proving the validity of an argument is called the *principle of demonstration*:

A demonstration that the argument $(p_1 \wedge p_2 \wedge \cdots \wedge p_n) \to q$ is valid is a sequence of propositions $s_1, s_2, \ldots, s_k$ such that s_k (the last proposition in the sequence) is q and each s_i, $1 \le i \le k$, in the sequence meets one or more of the following requirements:

a) s_i is one of the hypotheses.

b) s_i is a tautology.

c) s_i is a logical consequence of earlier propositions in the sequence.

Thus we see that under the assumption that the premises are true, each proposition in the demonstration will also be true and as the last proposition in the sequence is the conclusion of the argument, the demonstration shows (demonstrates) that if the premises are all true then the conclusion must also be true; i.e., the argument is valid.

As an example of this, let us consider the example above which we checked using a truth table:

$$\begin{array}{c} \neg q \\ \underline{p \to q} \\ \neg p. \end{array}$$

In writing out the demonstration it is helpful to the reader to include the justification for each proposition being in the sequence. We usually don't include the names and numbers of the tautologies used, but as an aid for beginners, they are included here.

Proposition	Reason
1. $\neg q$	hypothesis
2. $p \to q$	hypothesis
3. $\neg q \to \neg p$	contrapositive of 2 (13 on the tautology list)
4. $\neg p$	logical consequence of 1, 3 (19, modus ponens)

There are many ways of doing a demonstration correctly and even in this simple case we can proceed a little differently:

Proposition	Reason
1. $\neg q$	hypothesis
2. $p \rightarrow q$	hypothesis
3. $\neg p$	logical consequence of 1, 2 (20, modus tollens)

For a somewhat more complicated example, consider:

$$p \vee q$$
$$q \rightarrow \neg p$$
$$\underline{p \rightarrow q}$$
$$q.$$

Proposition	Reason
1. $q \rightarrow \neg p$	hypothesis
2. $p \rightarrow q$	hypothesis
3. $\neg q \rightarrow \neg p$	contrapositive of 2
4. $(q \vee \neg q) \rightarrow \neg p$	logical consequqnce of 1, 3 (15 a) on list)
5. $q \vee \neg q$	tautology
6. $\neg p$	logical consequence of 4, 5 (modus ponens)
7. $p \vee q$	hypothesis
8. q	logical consequence of 6, 7 (22, disjunctive syllogism)

For variety, here is another demonstration of the same argument (you might try to find some others yourself):

Proposition	Reason
1. $q \rightarrow \neg p$	hypothesis
2. $p \rightarrow q$	hypothesis
3. $p \rightarrow \neg q$	contrapositive of 1
4. $p \rightarrow (q \wedge \neg q)$	logical consequence of 2, 3 (15 b) on list)
5. $\neg p$	logical consequence of 4 (23, absurdity)
6. $p \vee q$	hypothesis
7. q	logical consequence of 5, 6 (disjunctive syllogism)

An extension of the principle of demonstration, called the method of *indirect proof* (or proof by contradiction), is based on the reductio ad absurdum logical equivalence (14 on the list). Applying this form to our argument we obtain

$$((p_1 \wedge p_2 \wedge \cdots \wedge p_n) \to q) \leftrightarrow ((p_1 \wedge p_2 \wedge \cdots \wedge p_n \wedge \neg q) \to \mathbf{c}).$$

Since this is a logical equivalence we can substitute the right-hand side for the left-hand side. This means, as far as our demonstration is concerned, that we have an additional hypothesis, $\neg q$ (the negation of the conclusion), and our demonstration will be complete when we obtain a contradiction (any contradiction).

As an example of this method, let us consider again the argument used in the previous example:

Proposition	Reason
1. $\neg q$	hypothesis (negation of conclusion in indirect proof)
2. $p \vee q$	hypothesis
3. p	logical consequence of 1, 2 (disjunctive syllogism)
4. $p \to q$	hypothesis
5. q	logical consequence of 3, 4 (modus ponens)
6. $q \wedge \neg q$	logical consequence of 1, 5 (this is the contradiction we were looking for)
7. q	logical consequence of 6 (indirect proof)

It is interesting to note that the hypothesis $q \to \neg p$ was not used in this proof, although it was in the two previous proofs. You might try to find a direct proof of the validity of the argument without using this hypothesis.

The principle of demonstration provides a good method of establishing the validity of arguments but it does have the disadvantage of not showing that an argument is invalid. The fact that we cannot give a demonstration of a particular argument is not sufficient to show that the argument is invalid; perhaps we are just not clever enough. However, there is another way, other than using truth tables, of showing an argument is invalid. If we recall what is meant by a valid argument, we will remember that the conclusion must be true whenever all the premises are true so if we can find just *one* case where the premises are true but the conclusion is false, then we have shown that the argument is invalid. Sometimes, in failing to obtain a demonstration, we are led to such a case, often called a *counterexample* to the argument. For example, consider the following argument:

$$p \to q$$
$$\underline{\neg p \vee q}$$
$$q \to p.$$

With a little thought we see that if q is T and p is F then the conclusion is F while both the premises are T; thus, the argument is invalid.

Exercises 1.5

1. Determine the validity of the following arguments using truth tables:

   a) $p \rightarrow q$ b) $p \vee q$ c) $p \vee \neg q$
       $\underline{\neg p \vee q}$ $\underline{r \rightarrow q}$ $\underline{\neg p}$
       $q \rightarrow p$ $\underline{q}$ $\neg q$
                $\neg r$

2. Give examples of the following where possible; if not possible, state why:
   a) An invalid argument with a false conclusion.
   b) A valid argument with a true conclusion.
   c) An invalid argument with a true conclusion.
   d) A valid argument with a false conclusion.
   e) A valid argument with true hypotheses and a false conclusion.
   f) An invalid argument with true hypotheses and a false conclusion.
   g) A valid argument with false hypotheses and a true conclusion.

3. Establish the validity of the following arguments using the principle of demonstration or show by counterexample that they are invalid:

   a) $\neg p \vee q$ b) $p \rightarrow q$ c) $\neg p \vee q$
       $\underline{p}$ $\underline{r \rightarrow \neg q}$ $\underline{\neg r \rightarrow \neg q}$
       q $p \rightarrow \neg r$ $p \rightarrow \neg r$

   d) $q \vee \neg p$ e) $\neg p$ f) $(p \wedge q) \rightarrow (r \wedge s)$
       $\underline{\neg q}$ $\underline{p \rightarrow q}$ $\underline{\neg r}$
       p $\neg p \vee \neg q$

   g) $p \rightarrow q$ h) $p \vee q$ i) $p \rightarrow q$
       $\neg q \rightarrow \neg r$ $q \rightarrow \neg r$ $\neg r \rightarrow \neg q$
       $s \rightarrow (p \vee r)$ $\underline{\neg r \rightarrow \neg p}$ $\underline{r \rightarrow \neg p}$
       $\underline{s}$ $\neg(p \wedge q)$ $\neg p$
       q

   j) $\underline{p \rightarrow \neg p}$ k) $p \vee q$ l) p
       $\neg p$ $p \rightarrow r$ $q \rightarrow \neg p$
                    $\underline{\neg r}$ $\neg q \rightarrow (r \vee \neg s)$
                    q $\underline{\neg r}$
                               $\neg s$

m) $p \to (q \lor s)$
$q \to r$

$\overline{p \to (r \lor s)}$

n) $p \to \neg q$
$q \to p$
$r \to p$

$\overline{\neg q}$

o) $p \to q$
$r \to s$
$\neg(p \to s)$

$\overline{q \land \neg r}$

1.6 QUANTIFIERS

When we were first discussing propositions we noted that "$x < 3$" was not a proposition since we did not know what x represented and thus could not assign a truth value. In this case we call x a *variable* (a symbol which may take on various values) and "$x < 3$" a *propositional function*. Actually this is a slight abuse of language since "$x < 3$" is really a propositional-valued function; that is, for each (properly chosen) value for x we get a proposition. This is similar to the real-valued functions we studied in precalculus (and perhaps elsewhere). For example, if f is the function given by $f(x) = 2x - 3$ then for each value of x in the domain of f (which we will take to be the set of real numbers) f returns a real number; i.e., $f(x)$ is a real number. Thus, $f(-1) = -5$, $f(5) = 7$. If we adopt a similar functional notation for "$x < 3$," say $p(x)$, and let the domain of p be the set of real numbers, then for each choice of x in the domain of p, $p(x)$ is a proposition. For example, when $x = 2$ we get $p(2)$ which is "$2 < 3$" and when $x = 8$ we get $p(8)$ or "$8 < 3$." Note that $p(2)$ is a true proposition while $p(8)$ is a false proposition.

Thus we will say that if r is a declarative sentence containing one or more variables and r becomes a proposition when particular values (sometimes called *interpretations*) are given to the variables, then r is a propositional function. As is the case with our real-valued functions from precalculus, the set of possible values for the variable is called the *domain* of the propositional function. Sometimes the domain will be explicitly stated and other times the domain will have to be inferred from the context. We will denote propositional functions by p, q, etc., and (as in the case of real-valued functions) use $p(x)$, $q(x, y)$ (to be read as "p of x," "q of x, y") to indicate defining "formulas" for these functions. Thus, if $p(x)$ is "$x < 3$" then $p(1)$, $p(-7)$, $p(0)$ are true while $p(3)$, $p(12)$, $p(\pi)$ are false; if $q(x, y)$ is "$x < y$" then $q(1, 2)$, $q(-2, 14)$ and $q(0, 5)$ are true while $q(0, 0)$, $q(2, 1)$ and $q(\pi, 3)$ are false.

Suppose that D is the domain of a propositional function p. We know that we can make p into a proposition by substituting various members of D into p; however, this is not the only way in which p can be made into a proposition. Another method is called *quantification* and there are two ways in which we quantify propositional functions. One is to preface the propositional function with "for all x in D" (or "for every x in D") and the other is to preface it with "there exists an x in D such that" (or "some x in D have the property that"). The notation we will use for this is

For all x in D, $p(x)$ is denoted by $\forall x$ in D, $p(x)$.

There exists an x in D such that $p(x)$ is denoted by $\exists x$ in $D \ni p(x)$.

$\forall$ is called the *universal* quantifier and is translated as "for all," $\exists$ is called the *existential* quantifier and is translated as "there exists" and $\ni$ is the symbol for "such that." We assign truth values to these propositions in accord with the usual meaning we give to "for all" and "there exists":

$$\forall x \text{ in } D, p(x)$$

will be given a truth value of true if $p(x)$ is true for *every* interpretation of x in D; otherwise it is given a truth value of false.

$$\exists x \text{ in } D \ni p(x)$$

will be given a truth value of true if $p(x)$ is true for *at least one* interpretation of x in D; otherwise it will be given a truth value of false. Thus, we see that if D is finite, say with elements $x_1, x_2, \ldots, x_n$, then

$$\forall x \text{ in } D, p(x)$$

is equivalent to a conjunction; namely

$$p(x_1) \wedge p(x_2) \wedge \cdots \wedge p(x_n),$$

while

$$\exists x \text{ in } D \ni p(x)$$

is equivalent to a disjunction, that is,

$$p(x_1) \vee p(x_2) \vee \cdots \vee p(x_n).$$

For example, if $D = \{1, 2, 3, 4\}$, $S = \{-1, 0, 1, 2\}$ and p is the propositional function given by $p(x)$ is "$x < 3$" then

$$\forall x \text{ in } D, p(x)$$

is false (since $p(3)$ is false), while

$$\forall x \text{ in } S, p(x); \quad \exists x \text{ in } D \ni p(x); \quad \exists x \text{ in } S \ni p(x)$$

are true. Note that the truth value of a quantified propositional function depends on the domain used. With p and S as above, let's look at this in another way.

$$\forall x \text{ in } S, p(x)$$

is equivalent to

$$p(-1) \wedge p(0) \wedge p(1) \wedge p(2),$$

while

$$\exists x \text{ in } S \ni p(x)$$

is equivalent to

$$p(-1) \vee p(0) \vee p(1) \vee p(2).$$

Thus, if you were a computer program (say) checking on the truth value of $\forall x$ in S, $p(x)$, you would have to take each element x in S and check the truth value of $p(x)$. As soon as you found a value of false you would return a value of false for $\forall x$ in S, $p(x)$; otherwise you would return a value of true after checking every element in S. Similarly, to determine the truth value of $\exists x$ in $S \ni p(x)$, you would take each element x in S in turn and check the truth value of $p(x)$. As soon as you found a true you would return true as the truth value of $\exists x$ in $S \ni p(x)$; otherwise you would return false after checking all elements of S.

With the above in mind, we should be able to consider the special (degenerate) case when the domain in question is empty (contains no elements). For example, what truth values should be assigned to the propositions "All mathematicians over three meters tall like chocolate" and "There exists a mathematician over three meters tall who likes chocolate"? If we let D be the set of mathematicians over three meters tall (an example of an empty set) and let $p(x)$ be "x likes chocolate" then these propositions become

$$\forall x \text{ in } D, p(x) \text{ and}$$

$$\exists x \text{ in } D \ni p(x).$$

For the first to be false we must produce a tall mathematician who does not like chocolate. Since there are no (sufficiently) tall mathematicians, we certainly cannot produce one who does not like chocolate; hence, the first proposition must be true. Similarly, for the second to be true we must produce a tall mathematician who likes chocolate. We cannot, so the second must be false. To summarize; if D is *empty* then *no matter what $p(x)$ is,*

$$\forall x \text{ in } D, p(x) \text{ is true and}$$

$$\exists x \text{ in } D \ni p(x) \text{ is false.}$$

You may not like this, but that's the way it is.

A little thought should reveal how to form the negations of quantified propositional functions. Consider $\forall x$ in D, $p(x)$. If this is a false proposition then $p(x)$ is not true for all interpretations of x in D; that is, there is at least one value of x in D such that $p(x)$ is false. Thus we see that:

$$\neg(\forall x \text{ in } D, p(x)) \leftrightarrow \exists x \text{ in } D \ni \neg p(x).$$

Using similar reasoning (the reader should try to supply this) we obtain

$$\neg(\exists x \text{ in } D \ni p(x)) \leftrightarrow \forall x \text{ in } D, \neg p(x).$$

If D is finite, these are just extensions of DeMorgan's laws; try it in an example to see what is going on.

To illustrate the negation of a quantified propositional function, consider

$$\forall x \text{ in } D, [p(x) \rightarrow q(x)].$$

Using these ideas we obtain as the negation

$$\exists x \text{ in } D \ni [p(x) \wedge \neg q(x)].$$

One of the main difficulties in dealing with quantified propositional functions given in a natural language (English, in our case) is ascertaining the correct logical form of these quantified statements. Of course, if we are given something like "There exists an integer such that its square is 9," it is easy to see that its form is

$$\exists x \text{ in } \mathbb{Z} \ni p(x),$$

where $\mathbb{Z}$ is the set of integers and $p(x)$ is "$x^2 = 9$." Unfortunately, in most cases the English rendering is not so straightforward and a correct translation into symbols (which shows clearly the logical form) requires an understanding of the meaning of the sentence; the translation cannot be done in some easily prescribed manner or according to a simple algorithm. Sometimes the quantification itself is not explicitly mentioned, but understood or implied. This is also true for the domain, even if the quantifier is present. For example, most mathematical definitions and theorems involve quantifiers; however, quite often these quantifiers are not apparent to the casual reader (of course, none of *our* readers approaches mathematics casually!). Thus, "If f is differentiable then f is continuous" really means "For all functions f (in some set of functions), if f is differentiable then f is continuous." It is usually a safe bet to assume that every theorem has a universal quantifier lurking about somewhere, expressed or implied.

In addition to finding the quantifiers, another problem which may arise is the determination of the correct form for a quantified propositional function. For example, "All logic students understand quantifiers" clearly involves the universal quantifier, but what is its correct form? If we let our domain D be the set of all students, $p(x)$ be "x is a logic student" and

$q(x)$ be "x understands quantifiers" then a possibility seems to be $\forall x$ in D, $p(x) \wedge q(x)$. But this means "Every student is a logic student and understands logic," not the meaning of the original proposition, for there may be some students who are not logic students. A correct rendering is: $\forall x$ in D, $p(x) \rightarrow q(x)$, which means, "For every student, if the student is a logic student then that student understands quantifiers." Similarly, we might be tempted to represent "Some logic students understand quantifiers" by $\exists x$ in $D \ni p(x) \rightarrow q(x)$. However, this is not correct, for there may be no logic students in our set of students, making $\exists x$ in $D \ni p(x) \rightarrow q(x)$ true while the given statement will be true only if there is at least one logic student who understands quantifiers. The given statement can be correctly rendered by $\exists x$ in $D \ni p(x) \wedge q(x)$, which means there is at least one student who is a logic student and who understands quantifiers. We should realize that these forms are somewhat domain-dependent for if we simplify things and restrict our domain to just the set of logic students (say D') then the first proposition becomes $\forall x$ in D', $q(x)$ and the second becomes $\exists x$ in $D' \ni q(x)$. To summarize:

$$\text{"every } p \text{ is a } q\text{"}$$

can be represented by.

$$\forall x \text{ in } D, p(x) \rightarrow q(x)$$

while

$$\text{"some } p \text{ is a } q\text{"}$$

can be given by

$$\exists x \text{ in } D \ni p(x) \wedge q(x),$$

(D being the domain).

One way to determine if you understand the natural language version of a quantified propositional function is to attempt to negate it. There are several possible ways to approach this sort of problem. The one which requires the least amount of experience is to translate the statement into symbolic form, use our well-known rules for the negation of propositions and quantified propositional functions and then translate the result back into English. After a sufficient amount of practice you should be able to negate some statements directly, but even with considerable experience it is helpful to use symbolic representations to clarify the structure.

As an example of this, suppose we wished to negate "All logic students understand quantifiers," considered above. With D, p and q as before, a symbolic representation is

$$\forall x \text{ in } D, p(x) \rightarrow q(x).$$

Proceeding with the negation, step-by-step,

$$\neg[\forall x \text{ in } D, p(x) \rightarrow q(x)]$$
$$\leftrightarrow \exists x \text{ in } D \ni \neg[p(x) \rightarrow q(x)]$$
$$\leftrightarrow \exists x \text{ in } D \ni p(x) \land \neg q(x).$$

Thus, a negation of "All logic students understand quantifiers" is "There exists a student who is a logic student and who does not understand quantifiers," or more in the style of the original proposition, "Some logic students do not understand quantifiers." As a further check of our understanding, we might ask, "What would make 'All logic students understand quantifiers,' false?" After a little reflection (we hope) we would answer, "There must be a logic student who does not understand quantifiers," which, of course, will be true when our negation "Some logic students do not understand quantifiers" is true.

Exercises 1.6

1. Translate the following into symbolic form, indicating appropriate choices for domains:
   a) There exists an integer x such that $4 = x + 2$.
   b) For all integers x, $4 = x + 2$.
   c) Every equilateral triangle is equiangular.
   d) All students like logic.
   e) Some students dislike logic.
   f) No man is an island.
   g) Everyone who understands logic likes it.
   h) Each person has a mother.
   i) Amongst all the integers there are some which are primes.
   j) Some integers are even and divisible by 3.
   k) Some integers are even or divisible by 3.
   l) All cyclic groups are abelian.
   m) At least one of the letters in *banana* is a vowel.
   n) One day next month is a Friday.
   o) $x^2 - 4 = 0$ has a positive solution.
   p) Every solution of $x^2 - 4 = 0$ is positive.
   q) No solution of $x^2 - 4 = 0$ is positive.
   r) One candidate will be the winner.
   s) Every element in set A is an element of set B.

2. Find an English negation for each of the propositions in exercise 1.
3. Let D be the set of natural numbers (that is, $D = \{1, 2, 3, 4, \ldots\}$), $p(x)$ be "x is even," $q(x)$ be "x is divisible by 3" and $r(x)$ be "x is divisible

by 4." For each of the following, express in English, determine its truth value and give an English negation.

a) $\forall x$ in $D, p(x)$.

b) $\forall x$ in $D, p(x) \vee q(x)$.

c) $\forall x$ in $D, p(x) \rightarrow q(x)$.

d) $\forall x$ in $D, p(x) \vee r(x)$.

e) $\forall x$ in $D, p(x) \wedge q(x)$.

f) $\exists x$ in $D \ni r(x)$.

g) $\exists x$ in $D \ni p(x) \wedge q(x)$.

h) $\exists x$ in $D \ni p(x) \rightarrow q(x)$.

i) $\exists x$ in $D \ni q(x) \rightarrow q(x + 1)$.

j) $\exists x$ in $D \ni p(x) \leftrightarrow q(x + 1)$.

k) $\forall x$ in $D, r(x) \rightarrow p(x)$.

l) $\forall x$ in $D, p(x) \rightarrow \neg q(x)$.

m) $\forall x$ in $D, p(x) \rightarrow p(x + 2)$.

n) $\forall x$ in $D, r(x) \rightarrow r(x + 4)$.

o) $\forall x$ in $D, q(x) \rightarrow q(x + 1)$.

4. For each of the propositions in exercise 3 give (if possible) an example of a domain D' such that the proposition has the opposite truth value to that which it had with D the set of natural numbers.

5. Are the following always, sometimes or never true (give examples of domains D and propositional functions p or reasons to justify your answers)?

a) $[\forall x$ in $D, p(x)] \rightarrow [\exists x$ in $D \ni p(x)]$.

b) $[\exists x$ in $D \ni p(x)] \rightarrow [\forall x$ in $D, p(x)]$.

c) $[\forall x$ in $D, \neg p(x)] \rightarrow \neg[\forall x$ in $D, p(x)]$.

d) $[\exists x$ in $D \ni \neg p(x)] \rightarrow \neg[\exists x$ in $D \ni p(x)]$.

e) $\neg[\forall x$ in $D, p(x)] \rightarrow [\forall x$ in $D, \neg p(x)]$.

f) $\neg[\exists x$ in $D \ni p(x)] \rightarrow [\exists x$ in $D \ni \neg p(x)]$.

1.7 MORE QUANTIFIERS

Many mathematical statements involve more than one quantifier. Some examples of such statements are "For every even integer n there exists an integer k such that $n = 2k$," "For every line l and every point p not on l there exists a line l' through p which is parallel to l," "For all y in B there exists an x in A such that $f(x) = y$," "For all x in the domain of f and for all $\epsilon > 0$ there exists a $\delta > 0$ such that $|x - c| < \delta$ implies $|f(x) - L| < \epsilon$," "For every x in G there exists an x' in G such that $xx' = e$." As might be expected, the difficulties which presented themselves when we considered one quantifier persist when we have more than one quantifier and, in addition, new difficulties arise, so we will have to be especially careful in our analysis of these higher-level quantifications.

Let's look first at the structure of a proposition involving two different quantifiers, say

$$\forall x \text{ in } S, \exists y \text{ in } T \ni p(x, y).$$

How are we to read this? As usual, we read from left to right so this means

$$\forall x \text{ in } S, [\exists y \text{ in } T \ni p(x, y)].$$

Thus, if $S = \{1, 2\}$ and $T = \{3, 4\}$ then we have (applying the universal quantifier first, as required):

$$[\exists y \text{ in } T \ni p(1, y)] \wedge [\exists y \text{ in } T \ni p(2, y)].$$

Now applying the existential quantifier:

$$[p(1, 3) \vee p(1, 4)] \wedge [p(2, 3) \vee p(2, 4)].$$

(Readers familiar with computer programming will see the resemblance between this and nested loops.)

In contrast, consider the same quantified propositional function with the order of the quantifiers reversed; that is,

$$\exists y \text{ in } T \ni \forall x \text{ in } S, p(x, y).$$

Proceeding in the same way we obtain

$$[\forall x \text{ in } S, p(x, 3)] \vee [\forall x \text{ in } S, p(x, 4)],$$

and hence,

$$[p(1, 3) \wedge p(2, 3)] \vee [p(1, 4) \wedge p(2, 4)].$$

Note that these are not equivalent; for example, if $p(1, 3)$ and $p(2, 4)$ are both true while $p(2, 3)$ and $p(1, 4)$ are both false then the first is true but the second is false.

As a slightly more concrete example of this, let $S = \{1, 2\}$ and $p(x, y)$ be "$x = y$." Then (the reader should provide the details)

$$\forall x \text{ in } S, \exists y \text{ in } S \ni p(x, y)$$

becomes

$$[\exists y \text{ in } S \ni 1 = y] \wedge [\exists y \text{ in } S \ni 2 = y]$$

which is

$$[1 = 1 \text{ or } 1 = 2] \quad \text{and} \quad [2 = 1 \text{ or } 2 = 2],$$

a true proposition, while

$$\exists y \text{ in } S \ni \forall x \text{ in } S, p(x, y)$$

is

$$[\forall x \text{ in } S, x = 1] \bigvee [\forall x \text{ in } S, x = 2]$$

or

$$[1 = 1 \text{ and } 2 = 1] \quad \text{or} \quad [1 = 2 \text{ and } 2 = 2],$$

a false proposition.

We note that if a proposition of the form

$$\forall x \text{ in } S, \exists y \text{ in } T \ni p(x, y)$$

is true then for each x in S there must be some y in T such that $p(x, y)$ is true; however, the choice of y may depend on the choice of x. On the other hand, for

$$\exists y \text{ in } T \ni \forall x \text{ in } S, p(x, y)$$

to be true there must be some y in T, say y_0, such that for this particular y_0, $p(x, y_0)$ is true for *every* choice of x in S.

It may be helpful to have a graphical way of looking at this. Suppose that $S = \{1, 2, 3, 4\}$ and $T = \{1, 2, 3\}$. We can display all twelve possible choices in a rectangular array as below, with the o indicating the possibilities.

$$
\begin{array}{c c c c c c}
 & 3 & \circ & \circ & \circ & \circ \\
T & 2 & \circ & \circ & \circ & \circ \\
 & 1 & \circ & \circ & \circ & \circ \\
\hline
 & & 1 & 2 & 3 & 4 \\
 & & & S & &
\end{array}
$$

As usual, we will represent the first set (S) along the horizontal axis and the second set (T) along the vertical axis. To make sure we understand how the coordinates are represented, the values are shown below:

$$
\begin{array}{c c c c c c}
 & 3 & (1,3) & (2,3) & (3,3) & (4,3) \\
T & 2 & (1,2) & (2,2) & (3,2) & (4,2) \\
 & 1 & (1,1) & (2,1) & (3,1) & (4,1) \\
\hline
 & & 1 & 2 & 3 & 4 \\
 & & & S & &
\end{array}
$$

Now suppose that $p(1, 1)$, $p(2, 3)$, $p(3, 2)$ and $p(4, 1)$ are true and for all other values of x and y, $p(x, y)$ is false (these true values are indicated by the rectangles in the figure below):

In terms of this picture we see that for

$$\forall x \text{ in } S, \exists y \text{ in } T \ni p(x, y)$$

to be true there must be at least one rectangle in every vertical column, while for

$$\exists y \text{ in } T \ni \forall x \text{ in } S, p(x, y)$$

to be true there must be an entire horizontal row of rectangles. Thus, for the example given, the first is true while the second is false. It should be clear that whenever the second is true (an entire row of rectangles) the first must also be true (at least one rectangle in every column).

For a more homey example of this, let S be the set of all persons and let $p(x, y)$ represent "y is the mother of x." Then $\forall x$ in S, $\exists y$ in $S \ni p(x, y)$ means that everyone has a mother while $\exists y$ in $S \ni \forall x$ in $S, p(x, y)$ means that there is a person who is the mother of everyone, clearly two different statements.

Next, let's try to understand another homey example: "For every dog on the sofa there is a flea in the carpet with the property that if the dog is black then the flea has bitten the dog." Some questions which we should be able to answer (if we understand the meaning of this statement) are "What is the negation of the statement?" "What can we say about its truth value if a) there are no black dogs on the sofa? b) one particular flea has bitten every dog? c) there is a black unbitten dog? d) there are no fleas in the carpet?" How might we go about answering these questions? If we can't do so immediately, a good way to start is to translate the proposition into symbolic form. Let S be the set of dogs on the sofa, C be the set of fleas in the carpet, $p(x)$ be "x is black," and $q(x, y)$ be "y has bitten x." Then the proposition is

$$\forall x \text{ in } S, \exists y \text{ in } C \ni p(x) \rightarrow q(x, y).$$

Now the negation can be dealt with in a straightforward, step-by-step manner:

$$\neg[\forall x \text{ in } S, \exists y \text{ in } C \ni p(x) \rightarrow q(x, y)]$$
$$\leftrightarrow \exists x \text{ in } S \ni \neg[\exists y \text{ in } C \ni p(x) \rightarrow q(x, y)]$$
$$\leftrightarrow \exists x \text{ in } S \ni \forall y \text{ in } C, \neg[p(x) \rightarrow q(x, y)]$$
$$\leftrightarrow \exists x \text{ in } S \ni \forall y \text{ in } C, p(x) \wedge \neg q(x, y).$$

Thus the negation, in English, is "There is a dog on the sofa such that for each flea in the carpet, the dog is black and the flea hasn't bitten the dog," or more conversationally, "There is a black, unbitten dog on the sofa." Now we should be able to answer the other questions which were asked above. In situation a) the proposition is true since there must be an unbitten black dog for it to be false; in situation b) it is true since $q(x, y)$ will be true for all dogs x; in situation c) it is false since the negation is true. The truth value in situation d) cannot be decided without more information. If there are some black dogs on the sofa then it is false; if there are no black dogs then it is true. This gives an example of the sort of questions we should be able to answer if we understand the meaning of such a quantified propositional function.

   With two quantifiers and two domains there are eight possible orders in which the quantifiers may occur. We have already noted that when the quantifiers are mixed (that is, one universal and one existential), the order is important:

$$\forall x \text{ in } S, \exists y \text{ in } T \ni p(x, y)$$

is not necessarily the same as

$$\exists y \text{ in } T \ni \forall x \text{ in } S, p(x, y).$$

If both quantifiers are the same we do have equivalence (this is because the connectives are all the same—$\vee$ for $\exists$ and $\wedge$ for $\forall$; just the order is different and we know that both $\vee$ and $\wedge$ commute); thus:

$$[\exists x \text{ in } S \ni \exists y \text{ in } T \ni p(x, y)] \leftrightarrow [\exists y \text{ in } T \ni \exists x \text{ in } S \ni p(x, y)]$$

and

$$[\forall x \text{ in } S, \forall y \text{ in } T, p(x, y)] \leftrightarrow [\forall y \text{ in } T, \forall x \text{ in } S, p(x, y)].$$

If the domain is the same for both quantifiers we often shorten these by writing

$$\forall x, y \text{ in } S, p(x, y) \quad \text{for} \quad \forall x \text{ in } S, \forall y \text{ in } S, p(x, y) \quad \text{and}$$

$$\exists x, y \text{ in } S \ni p(x, y) \quad \text{for} \quad \exists x \text{ in } S \ni \exists y \text{ in } S \ni p(x, y).$$

While the mixed forms are not equivalent, we can say that

$$[\exists y \text{ in } T \ni \forall x \text{ in } S, p(x, y)] \Rightarrow [\forall x \text{ in } S, \exists y \text{ in } T \ni p(x, y)].$$

This is because, as we observed above, if the left-hand side is true then there is at least one element of T, say y_0, which makes $p(x, y_0)$ true for all x in S so this y_0 may be used for each x in the right-hand side.

There is another set of difficulties which may arise, and that is distinguishing between, for example,

"Every integer is even or odd,"

and

"Every integer is even or every integer is odd."

It is easy to see (we hope), that these are not equivalent, since the first is true while the second is false. To help analyze the situation, let's put these propositions in symbolic form. If we let D be the set of integers, $p(x)$ be "x is even," and $q(x)$ be "x is odd," then the first is

$$\forall x \text{ in } D, [p(x) \vee q(x)],$$

while the second is

$$[\forall x \text{ in } D, p(x)] \vee [\forall x \text{ in } D, q(x)].$$

The reason that these two are not equivalent is essentially the same reason that we did not have equivalence in the case of mixed quantifiers; the $\forall$ involves "ands" and taken in conjunction with the "or" the order in which the interpretations occur changes the meaning. Using the same reasoning we might suspect that

$$\exists x \text{ in } D \ni [p(x) \wedge q(x)],$$

and

$$[\exists x \text{ in } D \ni p(x)] \wedge [\exists x \text{ in } D \ni q(x)]$$

are not equivalent; also, since $p \rightarrow q$ is equivalent to a disjunction $(\neg p \vee q)$ we would expect that

$$\forall x \text{ in } D, [p(x) \rightarrow q(x)]$$

and

$$[\forall x \text{ in } D, p(x)] \rightarrow [\forall x \text{ in } D, q(x)]$$

are not equivalent. Our suspicions are well-founded as none of these pairs is equivalent; however, in each pair there is one which implies the other, so we do have the following logical implications:

$$\left[[\forall x \text{ in } D, p(x)] \vee [\forall x \text{ in } D, q(x)] \right] \Rightarrow \forall x \text{ in } D, [p(x) \vee q(x)],$$

$$\exists x \text{ in } D \ni [p(x) \wedge q(x)] \Rightarrow \left[[\exists x \text{ in } D \ni p(x)] \wedge [\exists x \text{ in } D \ni q(x)] \right],$$

$$\forall x \text{ in } D, [p(x) \rightarrow q(x)] \Rightarrow \left[[\forall x \text{ in } D, p(x)] \rightarrow [\forall x \text{ in } D, q(x)] \right].$$

We should also suspect that the order of "$\forall$" and "$\wedge$" or "$\exists$" and "$\vee$" do not change the meaning and again we would be correct for

$$\left[[\forall x \text{ in } D, p(x)] \wedge [\forall x \text{ in } D, q(x)] \right] \Longleftrightarrow \forall x \text{ in } D, [p(x) \wedge q(x)]$$

and

$$\exists x \text{ in } D \ni [p(x) \vee q(x)] \Longleftrightarrow \left[[\exists x \text{ in } D \ni p(x)] \vee [\exists x \text{ in } D \ni q(x)] \right].$$

The ideas and methods of analysis we have used for statements involving two quantifiers can be extended to three (and more) quantifiers. Some examples of these have been included in the exercises.

Exercises 1.7

1. Translate the following into symbolic form, indicating appropriate choices for domains:
   a) For every even integer n there exists an integer k such that $n = 2k$.
   b) For every line l and every point p not on l there exists a line l' through p which is parallel to l.
   c) For all y in B there exists an x in A such that $f(x) = y$.
   d) For all x in the domain of f and for all $\epsilon > 0$ there exists a $\delta > 0$ such that $|x - c| < \delta$ implies $|f(x) - L| < \epsilon$.
   e) For every x in G there exists an x' in G such that $xx' = e$.
   f) If every integer is odd then every integer is even.
   g) Everybody loves somebody sometime.
   h) From amongst all the fleas in the carpet there is one for which there exists on every dog on the sofa a bite which that flea has made.
   i) For every integer n there exists another integer greater than $2n$.
   j) The sum of any two even integers is even.
   k) Every closed and bounded subset of $\mathbb{R}$ is compact.

2. Find an English negation for each of the propositions in exercise 1.

3. Let $p(x, y)$ represent "$x + 2 > y$" and let D be the set of natural numbers ($D = \{1, 2, 3, \ldots\}$). Write out in words and assign truth values to
   a) $\forall x \text{ in } D, \exists y \text{ in } D \ni p(x, y)$.
   b) $\exists x \text{ in } D \ni \forall y \text{ in } D, p(x, y)$.
   c) $\forall x \text{ in } D, \forall y \text{ in } D, p(x, y)$.
   d) $\exists x \text{ in } D \ni \exists y \text{ in } D \ni p(x, y)$.

e) $\forall y$ in $D, \exists x$ in $D \ni p(x, y)$.

f) $\exists y$ in $D \ni \forall x$ in $D, p(x, y)$.

4. Let $D = \{1, 2\}$, $p(x)$ be "x is even" and $q(x)$ be "x is odd." Write out in detail the following quantifications as conjunctions and disjunctions of interpretations (as was done at the beginning of this section):

a) $\forall x$ in $D, [p(x) \wedge q(x)]$.

b) $[\forall x$ in $D, p(x)] \wedge [\forall x$ in $D, q(x)]$.

c) $\forall x$ in $D, [p(x) \vee q(x)]$.

d) $[\forall x$ in $D, p(x)] \vee [\forall x$ in $D, q(x)]$.

e) $\exists x$ in $D \ni [p(x) \wedge q(x)]$.

f) $[\exists x$ in $D \ni p(x)] \wedge [\exists x$ in $D \ni q(x)]$.

g) $\exists x$ in $D \ni [p(x) \vee q(x)]$.

h) $[\exists x$ in $D \ni p(x)] \vee [\exists x$ in $D \ni q(x)]$.

i) $\exists x$ in $D \ni [p(x) \rightarrow q(x)]$.

j) $[\exists x$ in $D \ni p(x)] \rightarrow [\exists x$ in $D \ni q(x)]$.

5. Give some examples to show that the following logical implications are not logical equivalences:

a) $\big[[\forall x$ in $D, p(x)] \vee [\forall x$ in $D, q(x)]\big] \Rightarrow \forall x$ in $D, [p(x) \vee q(x)]$.

b) $\exists x$ in $D \ni [p(x) \wedge q(x)] \Rightarrow \big[[\exists x$ in $D \ni p(x)] \wedge [\exists x$ in $D \ni q(x)]\big]$.

c) $\forall x$ in $D, [p(x) \rightarrow q(x)] \Rightarrow \big[[\forall x$ in $D, p(x)] \rightarrow [\forall x$ in $D, q(x)]\big]$.

6. Determine what relationship (if any) exists between

$$\exists x \text{ in } D \ni [p(x) \rightarrow q(x)]$$

and

$$[\exists x \text{ in } D \ni p(x)] \rightarrow [\exists x \text{ in } D \ni q(x)].$$

Give reasons and examples to support your answer.

7. Show that the second logical equivalence in each of the following pairs can be obtained from the first by negation:

a)

$$[\exists x \text{ in } S \ni \exists y \text{ in } T \ni p(x, y)] \Longleftrightarrow [\exists y \text{ in } T \ni \exists x \text{ in } S \ni p(x, y)]$$

and

$$[\forall x \text{ in } S, \forall y \text{ in } T, p(x, y)] \Longleftrightarrow [\forall y \text{ in } T, \forall x \text{ in } S, p(x, y)]$$

b)

$$\big[[\forall x \text{ in } D, p(x)] \wedge [\forall x \text{ in } D, q(x)]\big] \Longleftrightarrow \forall x \text{ in } D, [p(x) \wedge q(x)]$$

and

$$\exists x \text{ in } D \ni [p(x) \vee q(x)] \Longleftrightarrow \big[[\exists x \text{ in } D \ni p(x)] \vee [\exists x \text{ in } D \ni q(x)]\big].$$

8. Consider the following proposition:

For every chicken in the coop and for every chair in the kitchen there is a frying pan in the cupboard such that if the chicken's egg is in the frying pan then the chicken is within two meters of the chair.
a) Translate this into symbolic form.
b) Express its negation in symbols and in English.
c) Give two examples of circumstances under which it would be true.
d) Give two examples of circumstances under which it would be false.

1.8   METHODS OF PROOF

Now that we have learned the basics of logic we need to put our ideas to use in proving theorems. Of course, as you have observed in reading mathematics books, most proofs are written in an informal manner rather than in the rather formal style we used in our demonstrations in section 1.5. But despite this obvious difference in style, the logical structure used is the same in each case: assuming the hypotheses are true, we write down a sequence of propositions which are logical consequences of what we have written previously, ending with the conclusion of the theorem. For example, consider the following theorem and proof:

Theorem: If m and n are even integers then $m + n$ is an even integer. (Recall that an integer n is even if and only if there exists an integer k such that $n = 2k$; n is odd if and only if there exists an integer k such that $n = 2k + 1$.)

Proof: Let m and n be even integers. Then there exist integers j, k such that $m = 2j, n = 2k$. Thus $m + n = 2j + 2k = 2(j + k)$. Therefore $m + n$ is even.   □

Here we have what is known as a *direct proof*: we started by assuming the hypothesis (m and n are even integers) and developed a sequence of logical consequences, ending with the conclusion ($m + n$ is even). We should note that there are some hidden quantifiers lurking just below the surface which deserve to be examined. A complete statement of the theorem would be "$\forall m, \forall n$, (m is an even integer and n is an even integer) $\rightarrow$ ($m + n$ is an even integer)." How was it that we proved this theorem by considering just two integers (m and n) when we were supposed to show the conclusion holds for *all* integers? Would it be any different if we had observed that 2 and 4 are even and their sum, 6, is also even? Yes, very definitely! This proof above contains an example of the use of "fixed but arbitrary" variables. Observing that $2 + 4 = 6$ and that 6 is even only shows that the theorem is true for these two numbers (there might be something special about 2 or 4 which makes the proof work); however, if we choose two even integers and assume nothing else about them then the same reasoning could be used for *any* pair of even integers, so the proof is general and holds for all even integers. Thus, the term "fixed but arbitrary": m and n are fixed—we can carry out calculations with them—but arbitrary—they have no properties which are not shared by all even integers.

There are two other commonly used methods of proof, both based on familiar (we hope) logical equivalences: the contrapositive and reductio ad absurdum equivalences. For convenience we list them here (recall that **c** represents a contradiction—a proposition which is always false):

$$(p \rightarrow q) \Longleftrightarrow (\neg q \rightarrow \neg p) \text{ contrapositive}$$

$$(p \rightarrow q) \Longleftrightarrow ((p \wedge \neg q) \rightarrow \mathbf{c}) \text{ reductio ad absurdum.}$$

Let's see what these tell us about proving theorems. Suppose that we are interested in proving a theorem, say $p \rightarrow q$. The contrapositive law tells us that this is logically equivalent to its contrapositive, $\neg q \rightarrow \neg p$. Thus, we could prove the theorem by assuming $\neg q$ and ending with $\neg p$; that is, we start with the negation of the conclusion of the theorem and end with the negation of the hypothesis. We will call such a proof a *contrapositive* proof. As an example, consider a contrapositive proof of the theorem above, where our starting point will now be $m + n$ is not even (the negation of the conclusion):

Proof: Suppose that m, n are integers and $m + n$ is not even, i.e., odd. Then there exists an integer k such that $m + n = 2k + 1$. Now m is either even or odd. If m is odd the proof is finished, so assume that m is even. Then there exists an integer j such that $m = 2j$. Thus

$$n = (m + n) - m = 2k + 1 - 2j = 2(k - j) + 1,$$

so n is odd and the proof is complete. □

There are several points in this contrapositive proof which bear scrutiny. To help sort things out, let us analyze the form of the theorem, neglecting the quantifiers. Let p represent "m is an even integer," q represent "n is an even integer" and r represent "$m + n$ is an even integer." Then the theorem is

$$(p \wedge q) \rightarrow r.$$

Thus the contrapositive is

$$\neg r \rightarrow \neg (p \wedge q).$$

We can use DeMorgan's law to obtain the logically equivalent form:

$$\neg r \rightarrow (\neg p \vee \neg q),$$

and this is the form used in the above proof. A translation of this into words would be "If $m + n$ is odd then m is odd or n is odd." Thus the contrapositive form of the theorem has a disjunction as a conclusion. Recall that a disjunction is true when at least one of its subpropositions is true, so to show that the conclusion is true we need to show that m is odd or n is odd. The proof above did this by saying that m is odd or even (recall that $p \vee \neg p$ is a tautology) and then considered both cases (an example of exhaustive analysis): if m is odd then "m is odd or n is odd" is true and we are done; if m is even then n is odd (a little work was required here) so "m is odd or n is odd" is still true, which completed the proof. This is the usual technique for showing that a disjunction is true; i.e., if one subproposition is true you are done so you assume one subproposition is false and show that the other subproposition must be true.

The method of proof based on the reductio ad absurdum equivalence is called the method of *indirect proof* or *proof by contradiction* and was discussed in section 1.5. Recall that it involves starting with an additional hypothesis, the negation of the conclusion, and is complete when a contradiction is obtained. As an example, here is an indirect proof of the above theorem:

Proof: Suppose that m and n are even integers and that $m + n$ is odd. Then there exist integers j, k such that $m = 2j$ and $m + n = 2k + 1$. Thus

$$n = (m + n) - m = 2k + 1 - 2j = 2(k - j) + 1.$$

Therefore n is both odd and even, a contradiction, which completes the proof. $\square$

Before analyzing this proof, let us make sure we understand the reductio ad absurdum equivalence. Recall that $p \wedge \neg q$ is the negation of $p \rightarrow q$ so the reductio ad absurdum equivalence is equivalent to

$$(p \rightarrow q) \Longleftrightarrow (\neg(p \rightarrow q) \rightarrow \mathbf{c})).$$

If a proposition implies a contradiction (remember that $\mathbf{c}$ here represents a contradiction) then that proposition must be false (absurdity, number 23 on the list in section 1.4). Thus, if $\neg(p \rightarrow q) \rightarrow \mathbf{c}$ is true, $\neg(p \rightarrow q)$ must be false, that is, $p \rightarrow q$ is true. What this tells us about the indirect proof is that instead of proving $p \rightarrow q$ we can show $(p \wedge \neg q) \rightarrow \mathbf{c}$; that is, show that the conjunction of the original hypotheses, p, and the negation of the conclusion, $\neg q$, lead to a contradiction.

Translating this for the above theorem, the form of the indirect proof is (using p, q, r as before):

$$(p \wedge q \wedge \neg r) \rightarrow \mathbf{c},$$

or in words, "m is an even integer and n is an even integer and $m + n$ is an odd integer implies a contradiction." The particular contradiction we obtained in this case was "n is even and n is odd (not even)," although any contradiction would have served as well. One of the advantages of the indirect proof is that it gives us an additional hypothesis with which to work and is particularly useful in proving the non-existence of mathematical objects.

To summarize the forms of our three methods of proof:

a) Direct proof: Assume hypotheses

.

· (body of proof)

.

Conclusion.

b) Contrapositive proof: Assume negation of conclusion

.

· (body of proof)

.

Negation of hypotheses.

c) Indirect proof: Assume hypotheses and negation of conclusion

.

· (body of proof)

.

Contradiction.

In each of these forms, "body of proof" represents the logical consequences which follow from the assumptions and lead to the "conclusion," whether it is the original conclusion, the negation of the hypotheses or a contradiction, as the case may be.

It is important to note that there is *no* proof form in which one assumes the conclusion just as there is *no* form in which one assumes the negation of the hypothesis.

If the theorem to be proven is in the form $p \leftrightarrow q$ then the proof can be broken up into two parts, one showing $p \rightarrow q$ and the other showing the converse, $q \rightarrow p$.

As was the case with the principle of demonstration, we usually can't use our proof techniques to show that a conjecture is false. Naturally, our failure to produce a proof of a conjecture's truth is not sufficient to guarantee its falsity, so we resort to counterexamples. If we had the conjecture:

"If x is an odd integer and y is an even integer then $x + y$ is even"

we could show that it is false by producing the counterexample $x = 3$, $y = 2$ and observing that x is odd, y is even and their sum, 5, is odd. Thus we have produced an example satisfying the hypotheses but not the conclusion.

It is important to realize that the process of producing written proofs of propositions consists of two parts: understanding the ideas that make the proof work and writing out the proof in a logical and intelligible manner. These two parts require different mental activities and it is the interaction of the creative insight needed on the one hand with the rigors of logic on the other that is one of the main attractions of mathematics.

When one reads a mathematics book, it is possible to get the impression that mathematics develops in a linear, logical fashion, each new result following on the heels of the one before. This is somewhat misleading, for the formal presentation of mathematics does not mirror the mental activity involved in its creation. There is much trial and error, consideration of examples, false starts and other such activities which take place behind the scenes before the final form of the proof emerges into public view. In fact, in trying to write out a proof, the actual writing part comes last, after an understanding is obtained of why the conclusion follows from the hypotheses. Thus, when asked to prove something, don't expect to start writing out the proof immediately; thought and understanding must come first.

Exercises 1.8

1. Show that you understand the correct form for direct, contrapositive and indirect proofs by writing out the first and last lines of such proofs for each of the following theorems:
   a) If m is an even integer then m^2 is even.
   b) If f is a differentiable function then f is a continuous function.
   c) L is a one-to-one linear transformation if and only if $\ker L = \{0\}$.
   d) If (a_n) is bounded and monotonic then (a_n) converges.
   e) The homomorphic image of a cyclic group is a cyclic group.
   f) If the only non-zero term in the p-adic expansion of n is a 1 then $n = p^k$ for some $k \geq 0$.
   g) If f is not continuous at c then $\lim_{x \to c} f(x)$ does not exist or $\lim_{x \to c} f(x) \neq f(c)$.

h) Every closed and bounded subset of $\mathbb{R}$ is compact.

i) If m is an integer of the form $2, 4, p^n, 2p^n$ where p is an odd prime and n is a positive integer then m has primitive roots.

2. Determine which of the following "proofs" are correct and which are incorrect. If a proof is correct, indicate the type and if a proof is incorrect, indicate why it is incorrect.

   Theorem: If x and y are even integers then $x - y$ is an even integer.

   a) "Proof 1": Suppose that x and y are both odd integers. Then there exist integers j, k such that $x = 2j + 1$ and $y = 2k + 1$. Thus

   $$x - y = 2j + 1 - (2k + 1) = 2(j - k)$$

   which is even.

   b) "Proof 2": Suppose that $x - y$ is even and x is odd. Then there exist integers j, k such that $x - y = 2j$ and $x = 2k + 1$. Thus

   $$y = y - x + x = -2j + 2k + 1 = 2(k - j) + 1$$

   so y is odd, a contradiction.

   c) "Proof 3": Suppose that $x - y$ is odd. Then there exists an integer j such that $x - y = 2j + 1$. If y is even we are finished, so suppose that y is odd, say $y = 2k + 1$ for some integer j. Thus

   $$x = x - y + y = 2j + 1 - (2k + 1) = 2(j - k)$$

   so x is even and the proof is completed.

   d) "Proof 4": Suppose that x is even and $x - y$ is also even. Then there exist integers j, k such that $x = 2j$ and $x - y = 2k$. Thus

   $$y = x - (x - y) = 2j - 2k = 2(j - k)$$

   so y is also even.

   e) "Proof 5": Suppose that x, y are even and $x - y$ is odd. Then there exist integers j and k such that $x = 2j$ and $y = 2k$. Now

   $$x - y = 2j - 2k = 2(j - k)$$

   so $x - y$ is even. But this contradicts our assumption that $x - y$ is odd so the proof is complete.

   f) "Proof 6": Suppose that $x - y$ is odd, say $x - y = 2j + 1$ for some integer j. If x is odd we are done, so assume that x is even, say $x = 2k$ for some integer k. Then

   $$y = x - (x - y) = 2k - (2j + 1) = 2(k - j) - 1 = 2(k - j - 1) + 1$$

   so y is odd and we are finished.

g) "Proof 7": Suppose that x and y are both even. Then there exist integers j, k such that $x = 2j$ and $y = 2k$. Thus

$$x - y = 2j - 2k = 2(j - k)$$

so $x - y$ is even.

h) "Proof 8": Suppose that $x - y$ is even. Then if x is odd we are done, so suppose that x is even. Then there exist integers j, k such that $x - y = 2j$ and $x = 2k$. Thus

$$y = x - (x - y) = 2k - 2j = 2(k - j)$$

so y is also even.

i) "Proof 9": Suppose that $x - y$ is odd, say $x - y = 2j + 1$ for some integer j. Then if x is odd, say $x = 2k + 1$ for some integer k, we have

$$y = x - (x - y) = 2k + 1 - (2j + 1) = 2(k - j)$$

so y is even and we are done.

j) "Proof 10": Suppose x and y are odd and $x - y$ is odd. Then there exits integers j, k such that $x = 2j + 1, y = 2k + 1$. Thus we have

$$x - y = 2j + 1 - (2k + 1) = 2(j - k)$$

so $x - y$ is both odd and even, a contradiction.

3. Give direct, contrapositive and indirect proofs (if possible) of
   a) If x is an even integer and y is an odd integer then $x + y$ is an odd integer.
   b) If x and y are odd integers then xy is an odd integer.

4. For the following conjectures, prove the true ones and give a counterexample for the false ones:
   a) If x is an integer and $4x$ is even then x is even.
   b) If x is an even integer then $4x$ is even.
   c) If x is an integer and x^2 is even then x is even.
   d) If x is an integer and $3x$ is even then x is even.
   e) If x, y, z are integers and $x + y + z$ is odd then an odd number of x, y, z is odd.

5. It would seem that there might be a fourth proof form; an indirect proof of the contrapositive of a theorem. Explain why this was not mentioned in the discussion above.

CHAPTER
2

SETS,
RELATIONS
AND FUNCTIONS

2.1 SETS

The theory of sets forms the basis of almost all of mathematics. Surprisingly, it is a relatively recent development, having been started by Georg Cantor, a German mathematician, in the 1870's [for a short history, the reader is referred to *A History of Set Theory* by Phillip E. Johnson, Prindle, Weber & Schmidt, Boston, 1972]. On the surface set theory appears to be quite simple, but some of the problems arising from it are very subtle. These problems are still the subject of study and debate among mathematicians and this study has led to a deeper understanding of the foundations of mathematics. Thus set theory has turned out to be one of the more fruitful ideas in all of mathematics.

While it is possible to develop set theory from an axiomatic point of view, an informal approach will be more suitable for our purposes [*Naive Set Theory* by Paul R. Halmos, Springer-Verlag, New York, 1974, provides a relatively easy-to-read introduction to a more axiomatic development]. To get started, we may think of a set as Cantor did, as "a collection of definite distinguishable objects, called elements, thought of as a whole." Obviously, this cannot serve as a definition for *set* unless *collection* has been defined previously and we quickly see that we are trapped in a circular pattern of definitions. In fact, in any language there must be terms which are undefined within that language. This is also true of mathematics and we will take both *set* and *element of a set* to be primitive, undefined terms.

From an intuitive point of view it would seem that any collection of objects could be considered as a set (certainly early on Cantor thought that this was so); however, this is not the case. Towards the end of the last century mathematicians discovered that allowing any collection of objects to be a set led to logical paradoxes of several sorts. Two main ways were found to eliminate these paradoxes; sets were not allowed to be "too large" or restrictions were put on the sort of objects which could be elements of sets [for an interesting discussion of the questions involved, see *Set Theory* by Charles C. Pinter, Addison-Wesley, Reading, 1971]. These difficulties need not concern us here, but we will make the assumption that all sets under discussion at a particular time consist of elements taken from a universal set, usually denoted by $\mathbb{U}$. [To see the sort of problems which can arise if this is not done, see exercise 11 at the end of this section.] Quite often $\mathbb{U}$ will not be mentioned explicitly, so this situation is very similar to the idea of the domain of a variable for propositional functions. Thus, when we write $\forall a$, a in $A \to a$ in B, we will really mean $\forall a$ in $\mathbb{U}$, a in $A \to a$ in B, for some universal set $\mathbb{U}$.

We will usually denote sets by uppercase letters A, B, C, etc., and elements of sets by lowercase letters a, b, c, etc. "a is an element of set A" (or "a is a member of A," "a is in A") can be denoted by

$$a \in A$$

and "a is not an element of set A" by

$$a \notin A.$$

If a set does not have many elements, it is convenient to list them, enclosed in braces: $\{\cdots\}$. Thus if A is the set with elements 1, 2, 3, 4 we will indicate this by

$$A = \{1, 2, 3, 4\}.$$

Another way of specifying the elements of a set is to give a rule for set membership. Hence, if A is as above we could also indicate this by

$$A = \{a : a \text{ is an integer and } 1 \leq a \leq 4\} \text{ or}$$
$$A = \{x : (x - 2)(x - 1)(x - 4)(x - 3) = 0\}.$$

The notation $\{a : p(a)\}$ is read as "the set of all a such that $p(a)$ is true" (here p is some propositional function). This can also be written as $\{a \mid p(a)\}$. Note that the order in which the elements of a set are listed makes no difference, for if the order were important then the method of specifying the elements of a set by giving the membership rule would not work, for such a specification need not imply a particular order.

The alert reader may have noticed that we have already used equality of sets without mentioning what is meant by "set A is equal to set B." To remedy this deficiency, we make the following definition:

Definition 2.1: Set A *is equal to* set B, denoted by $A = B$, if and only if every element of A is an element of B and every element of B is an element of A. In symbols,

$$(A = B) \leftrightarrow [(\forall x, x \in A \to x \in B) \wedge (\forall x, x \in B \to x \in A)] \text{ or}$$

$$(A = B) \leftrightarrow (\forall x, x \in A \leftrightarrow x \in B).$$

Thus two sets are equal if and only if they have the same elements; e.g.,

$$\{1, 2, 3\} = \{2, 3, 1\} = \{x : 1 \le x \le 3 \text{ and } x \text{ is an integer}\}.$$

There are several sets which are used so frequently in mathematics that we give them special names; perhaps these are familiar to you:

$$\mathbb{N} = \{x : x \text{ is an integer and } x \ge 1\}$$

$$= \{1, 2, 3, 4, \ldots\} \text{ (the set of natural numbers)}$$

$$\mathbb{Z} = \{x : x \text{ is an integer}\}$$

$$= \{\ldots, -2, -1, 0, 1, 2, \ldots\} \text{ (the set of integers)}$$

$$\mathbb{Q} = \{\frac{x}{y} : x, y \in \mathbb{Z}, y \ne 0\}$$

$$= \{\ldots, \frac{-4}{3}, \frac{-3}{2}, \frac{-2}{1}, \frac{-1}{1}, \frac{0}{2}, \frac{1}{3}, \ldots\} \text{ (the set of rational numbers)}$$

$$\mathbb{R} = \{x : x \text{ is a real number}\} \text{ (the set of real numbers,}$$
$$\text{the points on the number line)}$$

Suppose that $A = \{1, 2, 3\}$ and $B = \{1, 2, 3, 4\}$. We easily see that $A \ne B$ ($4 \in B$ but $4 \notin A$) but we do notice that every element of A is also an element of B. This is an important concept so we give it a name:

Definition 2.2: Let A, B be sets. We say that A is a *subset* of B (or equivalently, B is a *superset* of A) if and only if every element of A is an element of B. This is denoted by

$$A \subseteq B \text{ or } B \supseteq A.$$

In symbols,

$$A \subseteq B \leftrightarrow (\forall x, x \in A \rightarrow x \in B).$$

If A is not a subset of B we will write $A \nsubseteq B$. Applying our techniques for negating quantified propositional functions we obtain

$$A \nsubseteq B \leftrightarrow \neg(\forall x, x \in A \rightarrow x \in B) \leftrightarrow (\exists x \ni x \in A \wedge x \notin B).$$

Note that for any set A, it is true that $A \subseteq A$. If $A \subseteq B$ but $A \neq B$ we will say that A is a *proper* subset of B and write

$$A \subset B \text{ or } B \supset A.$$

Similarly, if A is not a proper subset of B we write $A \not\subset B$. [Note that some authors do not make this distinction and use $\subset$ as we have used $\subseteq$ and thus have no distinct symbol for proper subset.]

As examples of this notation, if $A = \{1, 2, 3, 4\}, B = \{x : (x - 2)(x - 1)(x - 4)(x - 3) = 0\}$ and $C = \{1, 2, 3, 4, 5, 6\}$ then

$$A \subseteq B, B \subseteq C, C \subseteq \mathbb{N}$$

and

$$A \not\subset B, B \subset C, C \subset \mathbb{N}.$$

In letting a set be a collection of objects, obviously it is possible to have a collection with no objects: for example, the set of all mathematics students who are more than six meters tall. We call a set with no elements an *empty* set. If we have two such sets, say the set described above and the set of all mathematics instructors who are less than six millimeters tall, then we should see that these sets are equal (exercise 4) since they satisfy the definition of set equality. Thus, as any two such sets will be the same (equal), there is only one empty set and we are justified in speaking of *the* empty set and can represent it by a symbol, $\varnothing$. In symbols,

$$\varnothing = \{x : p(x) \wedge \neg p(x)\},$$

where p is any propositional function.

Since "$\forall x \in D, x \in \varnothing$" is false if D is non-empty, "$\forall x \in D, x \in \varnothing \rightarrow$ (any propositional function)" is true (sometimes we say that such an implication is *vacuously* true as the premise is never true) so in particular,

$$\forall x, x \in \varnothing \rightarrow x \in A$$

is true for any set A. Therefore, for any set A, we have $\varnothing \subseteq A$.

If one is not thinking carefully, it may be possible to confuse the ideas of set membership and subset. To help keep these ideas distinct, remember that "subsetness," that is, "is a subset of," is a relation which exists (or not) between *sets*, while "membership," that is, "is an element of," is a relation which exists (or not) between elements and sets. Thus, for a statement involving $\subseteq$ to make sense there must be sets on both sides of $\subseteq$, while an element should be to the left of $\in$ and a set which contains elements of that sort to the right. For example, all these propositions are true:

$$1 \in \{1, 2\}, 1 \not\subseteq \{1, 2\}, \{1\} \subseteq \{1, 2\}, \varnothing \notin \{1, 2\}, \varnothing \subseteq \{1, 2\}$$

while these are false:

$$1 \subseteq \{1, 2\}, 1 \notin \{1, 2\}, \{1\} \in \{1, 2\}, \varnothing \in \{1, 2\}.$$

Following the pattern in logic where we combined propositions using connectives to obtain new propositions, we wish to combine sets to obtain new sets using what we will call *set operations*. Their definitions are

---

Definition 2.3: Let A, B be sets. Then the *union* of A and B (denoted by $A \cup B$) is the set of all elements which are in at least one of A or B. In symbols,

$$A \cup B = \{x : x \in A \lor x \in B\}.$$

The *intersection* of A and B (denoted by $A \cap B$) is the set of all elements which are in both A and B. Thus,

$$A \cap B = \{x : x \in A \land x \in B\}.$$

If $A \cap B = \varnothing$, we say that A and B are *disjoint*. The *relative complement* of A in B (or the complement of A with respect to B), denoted by $B - A$ (sometimes by $B \setminus A$) is the set of all elements in B which are *not* in A. In symbols,

$$B - A = \{x : x \in B \land x \notin A\}.$$

If B is $\cup$, the universal set, then $\cup - A = \{x : x \in \cup \land x \notin A\} = \{x : x \notin A\}$ is called the *complement* of A and is denoted by A^C.

A convenient way of displaying sets so that we may have a visual image of unions, intersections and relative complements is to use a *Venn diagram* (see below for examples). In this sort of diagram we use regions to represent sets; the region enclosed by the rectangle represents the universal set and the regions inside the circles represent the sets indicated (and hence the region outside the circle A represents A^C). Thus the region shared by the two circles represents $A \cap B$ (shaded in the following diagram):

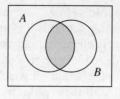

$$A \cap B$$

while the two regions for A and B taken together represent $A \cup B$

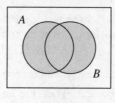

$$A \cup B$$

and the region shaded below represents $A - B$

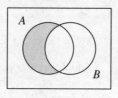

$$A - B$$

The same principle can be used when three sets are involved, for example (you should check this), the shaded region below represents $A \cap (B \cup C)$:

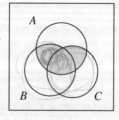

$$A \cap (B \cup C)$$

Note that when we had two sets the Venn diagram consisted of four regions while when we had three sets there were (count them!) eight regions. A little thought should show us why this is the case: for an element in $\mathbb{U}$ and each set involved, exactly one of two possibilities must hold: the element is in the set or it is not. Thus with two sets we have $2 \times 2 = 4$ possibilities while with three sets we have $2 \times 2 \times 2 = 8$ possibilities. You have probably never seen a Venn diagram for six sets; such a diagram would need $2^6 = 64$ regions and its complexity would make its usefulness limited indeed.

There is one other set obtained from a given set to which we give a name: namely, the set of all subsets of the given set.

Definition 2.4: Let A be a set. Then the set of all subsets of A, denoted by $\mathbb{P}(A)$ (or 2^A) is called the *power set* of A. In symbols

$$\mathbb{P}(A) = \{B : B \subseteq A\}.$$

To make sure these ideas are clear, let us consider a few examples:

Let $\mathbb{U}$ be the set of natural numbers (usually denoted by $\mathbb{N}$); i.e.,

$$\mathbb{U} = \mathbb{N} = \{x : x \text{ is an integer and } x \geq 1\}$$
$$= \{1, 2, 3, 4, \ldots\}.$$

We define

$$A = \{x : x \text{ is even}\},$$
$$B = \{x : x = 2k - 1 \text{ for some } k \in \mathbb{N}\},$$
$$C = \{y : y \leq 4\},$$
$$D = \{1, 3\}.$$

Then (you should verify these):

$$A \cup B = \mathbb{U},$$
$$A \cap B = \varnothing,$$
$$A \text{ and } D \text{ are disjoint,}$$
$$A^C = B,$$
$$B^C = A,$$
$$A - B = A,$$
$$C \cap D = D,$$
$$\mathbb{P}(D) = \{\varnothing, \{1\}, \{3\}, \{1, 3\}\},$$

$$C \nsubseteq D,$$
$$D - C = \varnothing,$$
$$D \subseteq C,$$
$$D \subset C,$$
$$D^C \supseteq A,$$
$$1 \nsubseteq D,$$
$$1 \in D,$$
$$A \cup C = A \cup D,$$
$$\varnothing \in \mathbb{P}(D),$$
$$\varnothing \subseteq \mathbb{P}(D),$$
$$\{1\} \in \mathbb{P}(D),$$
$$1 \notin \mathbb{P}(D).$$

Next we wish to prove some theorems involving sets, but first it will be helpful to make some observations about such proofs. One of the more common types of proof is a direct proof to show that one set is a subset of another set. Here is a specific example:

Theorem: Let A and B be sets with $A \cap B = A$. Then $A \subseteq B$.

Proof: Suppose A, B are sets with $A \cap B = A$. Let $a \in A$. Then

$\cdots$

"something or other using the assumption $A \cap B = A$"

$\cdots$

so $a \in B$; thus $A \subseteq B$. $\qquad\qquad\qquad\qquad\qquad\qquad\qquad\qquad\qquad\qquad\square$

Some comments about the form of this proof are in order. Note that we started the proof by "setting the stage"; that is, we stated what our symbols represent and what assumptions we made about them. This is a good way to start any proof, although you may notice that proofs in mathematics books, especially upper-level books, often omit this and assume that the reader can infer from the context what the symbols represent and what assumptions are being made about them. This points out an aspect of proof-writing which requires judgement: how much detail should be included. There is no correct answer to this question, but a good rule-of-thumb is to include sufficient detail so that a person at the lower end of the assumed readership will be able to comprehend the proof. When you are first starting out you might aim for sufficient detail so that you could understand your own proof yourself next week—that is, after some time has elapsed and the ideas of the proof are not in the forefront of your mind. When in doubt, err on the side of more detail.

After we had set the stage, the proof began, "Let $a \in A$." This is another example of the use of a "fixed-but-arbitrary" variable. We assume that a is an element of A but nothing else about it is assumed. But wait; there is more here than meets the (undiscerning) eye! We are actually considering *two* cases, one of which is not even mentioned. When we say, "let $x \in A$," we are assuming that $A \neq \varnothing$. What if $A = \varnothing$? This is the unmentioned case which has been swept under the rug. The reason that we leave this case unmentioned is that if $A = \varnothing$, we are done; for $\varnothing$ is a subset of any set, in particular B. Another more general way of thinking about this is to realize that the proposition we are trying to prove ($\forall x, x \in A \rightarrow x \in B$) is a universally quantified implication. If there are no elements in A, the implication is vacuously true. If we were to write this out in detail (which usually is not done), we would begin with: "We consider the two cases: $A = \varnothing$ and $A \neq \varnothing$. If $A = \varnothing$ then $A \subseteq B$ and we are done. If $A \neq \varnothing$, let $a \in A \ldots$." The point to remember is that whenever you write something like "let $a \in A$," you should be sure that the case $A = \varnothing$ causes no problems. Another important point to notice is that since we must show that the conclusion of the theorem ($A \subseteq B$ or $\forall x, x \in A \rightarrow x \in B$) is true, our choice of a fixed-but-arbitrary element is determined by the *conclusion* rather than the hypotheses (which in this case included $A \cap B = A$). Quite often the "something or other" in the body of the proof turns out to be just translation of definitions; e.g., $x \in A \cap B$ implies $x \in A$ and $x \in B$. The "$\square$" at the end is a symbol to indicate that the proof has been completed.

We can use this same technique twice to show that two sets are equal; this is the case since $A = B$ if and only if $A \subseteq B$ and $B \subseteq A$. Therefore, to show $A = B$ we show $A \subseteq B$ and then show $B \subseteq A$. As an example of this sort of proof, consider:

Theorem 2.1: Let A and B be sets. Then $A - B = A \cap B^C$.

Proof: Suppose that A and B are sets. First we show that $A - B \subseteq A \cap B^C$. Let $x \in A - B$. Then $x \in A$ and $x \notin B$ (this is just the definition of $A - B$). But $x \notin B$ implies that $x \in B^C$. Therefore $x \in A$ and $x \in B^C$ so we have $x \in A \cap B^C$ and $A - B \subseteq A \cap B^C$.

Now, suppose that $x \in A \cap B^C$. This means that $x \in A$ and $x \in B^C$ (again, using the definition of the set in question, in this case an intersection). But $x \in B^C$ means that $x \notin B$. Therefore, $x \in A$ and $x \notin B$ or $x \in A - B$, so $A \cap B^C \subseteq A - B$.

Since we have shown that $A - B \subseteq A \cap B^C$ and $A \cap B^C \subseteq A - B$, we have shown that $A - B = A \cap B^C$. $\qquad\qquad\square$

We have included more detail than normal here as this is our first proof about sets. Usually the parenthetical remarks above would be left out and less explanation would be provided.

Here is another example, with somewhat less explanation—see if you can follow the argument:

Theorem 2.2: If A, B, C are sets with $A \subseteq B$ and $B \subseteq C$ then $A \subseteq C$.

Proof: Suppose that A, B, C are sets with $A \subseteq B$ and $B \subseteq C$. Let $a \in A$. Then since $A \subseteq B$ we have $a \in B$. Further, since $B \subseteq C$ and $a \in B$ we also have $a \in C$. Therefore, $A \subseteq C$. $\qquad\qquad\square$

Observe that the conclusion, $A \subseteq C$, determined our starting point; we needed to show that every element in A was also in C so we started with a fixed-but-arbitrary element $a \in A$ which we ultimately showed was in C.

For a little more complicated example consider:

Theorem 2.3: Let A, B be sets. Then $A \subseteq B \leftrightarrow A \cap B = A$.

Proof: Suppose that A and B are sets. First, to show that $A \subseteq B$ implies $A \cap B = A$, suppose that $A \subseteq B$. Let $z \in A \cap B$. Then $z \in A$ and $z \in B$. Thus $z \in A$ so $A \cap B \subseteq A$. Now let $z \in A$. Since $A \subseteq B$, $z \in B$ so we have $z \in A$ and $z \in B$ which means that $z \in A \cap B$. Thus we have also shown that $A \subseteq A \cap B$ which together with our previously proven $A \cap B \subseteq A$ implies $A = A \cap B$.

Now to show that $A \cap B = A$ implies $A \subseteq B$, assume $A \cap B = A$. Let $a \in A$. Then, since $A = A \cap B$, $a \in A \cap B$ so $a \in B$. But this implies that $A \subseteq B$. $\qquad\qquad\square$

There are several points of this proof that deserve comment. First we note that the basic form of the theorem is that of an equivalence, which means that the proof will probably involve showing that two implications are true. The one we started with was $A \subseteq B \rightarrow A \cap B = A$. Now recall that it is the *conclusion* which determines the form and starting place of the proof and here the conclusion is that two sets are equal, which we know generally requires two parts; that is, show $A \cap B \subseteq A$ and $A \subseteq A \cap B$. Also note that the hypothesis $A \subseteq B$ was only used in one of these (in showing that $A \subseteq A \cap B$) and was *not* used as the starting point of the proof. The second implication, $A \cap B = A \rightarrow A \subseteq B$, has as its conclusion $A \subseteq B$ so we used our usual subset-proving technique for it. For a little variety, we could have used an indirect proof for this part:

Suppose that $A \cap B = A$ and $A \nsubseteq B$. Then there exists an element a such that $a \in A$ and $a \notin B$. But $a \notin B$ means that $a \notin A \cap B$. Since $A \cap B = A$ this implies $a \notin A$, a contradiction. Therefore $A \subseteq B$.

For one last example, we include a proof that a certain set is empty; such proofs usually are done indirectly.

Theorem 2.4: Let A, B be sets. Then $A \cap (B - A) = \varnothing$.

Proof: Suppose that A and B are sets. As $\varnothing$ is a subset of any set we have $\varnothing \subseteq A \cap (B - A)$ so all we have to show is that $A \cap (B - A) \subseteq \varnothing$. We will do this indirectly, which means that we will assume that there exists an element in $A \cap (B - A)$ which is not an element of $\varnothing$ and get a contradiction (as there are no elements in $\varnothing$, this amounts to showing that assuming an element in $A \cap (B - A)$ leads to a contradiction). Suppose that there exists a $y \in A \cap (B - A)$. Then $y \in A$ and $y \in B - A$. But $y \in B - A$ implies that $y \in B$ and $y \notin A$. Thus we have $y \in A$ and $y \notin A$, a contradiction, which completes the proof. $\square$

The above is fairly typical of a proof showing that a certain set, say C, is empty in that it is of the form: $x \in C \rightarrow$ a contradiction. This is the method usually used.

Now you can get a chance to try your hand at some proofs, a fine opportunity to use your knowledge of both logic and sets!

Exercises 2.1

1. Let

$$\mathbb{U} = \{1, 2, 3, 4, 5, 6, 7, 8\},$$

$$A = \{1, 2, 3, 4\},$$

$$B = \{x : (x - 2)^2(x - 3) = 0\},$$

$$C = \{x : x \text{ is odd}\}.$$

Find:

a) $A \cup B$.

b) $A \cap (B \cup C)$.

c) $C - A$.

d) $C \cup A^C$.

e) $(A \cup C)^C$.

f) $A^C \cap C^C$.

g) $\mathbb{P}(B)$.

2. Write out an English version of the negation of $A \subseteq B$ given in this section.

3. Let $\mathbb{U} = \mathbb{R}$, the set of real numbers. Recall the interval notation used in precalculus:

$$(a, b) = \{x : a < x < b\},$$

$$(a, b] = \{x : a < x \leq b\},$$

$$[a, b) = \{x : a \leq x < b\},$$

$$[a, b] = \{x : a \leq x \leq b\},$$

$$(-\infty, a) = \{x : x < a\},$$

$$(-\infty, a] = \{x : x \leq a\},$$

$$(a, \infty) = \{x : a < x\},$$

$$[a, \infty) = \{x : a \leq x\}.$$

Find the following:

a) $[1, 3] \cap (2, 4)$.

b) $(-\infty, 2) \cap [-1, 0]$.

c) $(-\infty, 2) \cap [-1, 3]$.

d) $[0, 10] \cup (1, 11)$.

e) $(0, \infty) \cap (-\infty, 1)$.

f) $(1, \infty) \cap (-\infty, 0)$.

g) $[-2, 0] \cup [0, 2]$.

h) $[-2, 0] \cup (0, 2]$.

i) $[-2, 0) \cup (0, 2]$.

j) $[-2, 0] \cup [2, 0]$.

k) $(0, 4]^C$.

l) $\mathbb{P}([1, 1])$.

m) $\mathbb{P}([0, 1])$.

4. Show that the two empty sets mentioned in the discussion of the empty set are equal.

5. Suppose that A, B, C are sets and $\mathbb{U}$ is the universal set. Prove the following:

a) $A \cup \varnothing = A$.

b) $A \cap \varnothing = \varnothing$.

c) $A - \varnothing = A$.

d) $A \cup \mathbb{U} = \mathbb{U}$.

e) $A \cap \mathbb{U} = A$.

f) $A \cup A^C = \mathbb{U}$.

g) $A \cap A^C = \varnothing$.

h) $A - A = \varnothing$.

i) $A - B \subseteq A$.

j) $A \cap B \subseteq A$.

k) $A \cup B \supseteq A$.

l) $A \cap B \subseteq A \cup B$.

m) $(A^C)^C = A$.

n) $(A \cup B)^C = A^C \cap B^C$.

o) $(A \cap B)^C = A^C \cup B^C$.

p) $A \cup (B - A) = A \cup B$.

q) $(A \cup B) - (A \cap B) = (A - B) \cup (B - A)$.

r) $A - (B \cup C) = (A - B) \cap (A - C)$.

s) $A \cup (B \cap C) = (A \cup B) \cap (A \cup C)$.

t) $A \cap (B \cup C) = (A \cap B) \cup (A \cap C)$.

6. Suppose that A, B, C, D are sets and $\mathbb{U}$ is the universal set. For each of the following theorems state the hypotheses and conclusion and indicate the form of a direct proof. Then write out a proof of each.

a) $A \subseteq \varnothing \leftrightarrow A = \varnothing$.

b) $A \subset B \wedge B \subset C \to A \subset C$.

c) $A \subseteq B \leftrightarrow A \cup B = B$.

d) $A \subseteq B \leftrightarrow \mathbb{P}(A) \subseteq \mathbb{P}(B)$.

e) $A \subseteq B^C \leftrightarrow A \cap B = \varnothing$.

f) $(A \cup B = C \wedge A \cap B = \varnothing) \to B = C - A$.

g) $(A \subseteq C \wedge B \subseteq C) \leftrightarrow A \cup B \subseteq C$.

h) $(A \subseteq C \wedge B \subseteq D) \to (A \cup B \subseteq C \cup D)$.

i) $[(A \cap C = A \cap B) \wedge (A \cup C = A \cup B)] \to B = C$.

j) $A \subseteq B \leftrightarrow A^C \cup B = \mathbb{U}$.

k) $A - B \subseteq B \leftrightarrow A \subseteq B$.

l) $A \cap B = \mathbb{U} \leftrightarrow A = B = \mathbb{U}$.

m) $A \cup B \neq \varnothing \leftrightarrow A \neq \varnothing \vee B \neq \varnothing$.

n) $\mathbb{P}(A) = \mathbb{P}(B) \to A = B$.

7. *Believe It or Not:* Instructions. These exercises appear throughout the remainder of the text. A conjecture is given, followed by a "proof" which purports to show that the conjecture is true and a "counterexample" which purports to show that it is false. Your task is to separate the wheat from the chaff and determine what is correct, keeping in mind the possibility that all three are incorrect. A complete solution involves pointing out any errors present (at least one part must be wrong) and correctly disposing of the conjecture; i.e., proving it if it is true or giving a suitable counterexample if it is false.

Conjecture: Let A, B be sets with $A \subseteq B$. Then $A - B = \varnothing$.

"Proof": Suppose that A, B are sets with $A \subseteq B$. Let $x \in A - B$. Then $x \in B$ and $x \notin A$. But $A \subseteq B$ so $x \notin A$ implies $x \notin B$, a contradiction. Thus $A - B = \varnothing$. □

"Counterexample": Let $A = \{1, 2, 3\}, B = \{2, 3\}$. Then $A \subseteq B$ but $A - B \neq \varnothing$.

8. **Believe It or Not:** Conjecture: Let A, B, C, D be sets with $A \subset C$ and $B \subset D$. Then $A \cup B \subset C \cup D$.

"Proof": Suppose that A, B, C, D are sets with $A \subset C$ and $B \subset D$. Let $x \in A \cup B$. Then $x \in A$ or $x \in B$. Suppose $x \in A$. Then since $A \subset C$, $x \in C$. Thus $x \in C \cup D$. If $x \in B$, we also obtain $x \in C \cup D$ since $B \subset D$. Therefore $A \cup B \subset C \cup D$. □

"Counterexample": Let $A = \{1\}, B = \{2\}, C = D = \{1, 2\}$. Then $A \subset B, C \subset D$ but $A \cup B \not\subset C \cup D$.

9. With the Believe It or Not exercises there are eight possibilities: the conjecture is true or false, the proof is correct or not and the counterexample is correct or not. Which of these eight possibilities cannot occur?

10. Suppose that A, B and C are sets. Show the following by using some of the results of exercises 5 and 6 rather than our usual method of going back to the definitions:
    a) $A \subseteq B$ implies $A \cap B^C = \varnothing$.
    b) $A \cup (A \cap B) = A$.
    c) $A \cap (A^C \cup B) = A \cap B$.
    d) $A \cap C = \varnothing$ implies $A \cap (B \cup C) = A \cap B$.
    e) $A \subseteq B$ implies $A = B - (B - A)$.

11. Suppose that any collection of objects could be a set. Then we could have the "set of all sets." Consider the subset S of the set of all sets given by

$$S = \{A : A \notin A\}.$$

Thus S is the set of all sets which are not elements of themselves.
    a) Give examples of two sets which are elements of S.
    b) Give examples of two sets which are not elements of S.
    c) Show that $S \notin S$.
    d) Show that $S \notin S^C$.

We note that if any collection of objects could be a set then exactly one of c) or d) would be true. That neither is true is called *Russell's paradox*, after Bertrand Russell, an English mathematican and philosopher who discovered it in the early days of set theory.

12. Let A and B be sets. Consider the following conjectures. Prove the true ones and give counterexamples for the false ones.
    a) $\mathbb{P}(A) \cup \mathbb{P}(B) \subseteq \mathbb{P}(A \cup B)$.
    b) $\mathbb{P}(A) \cap \mathbb{P}(B) \subseteq \mathbb{P}(A \cap B)$.
    c) $\mathbb{P}(A \cup B) \subseteq \mathbb{P}(A) \cup \mathbb{P}(B)$.
    d) $\mathbb{P}(A \cap B) \subseteq \mathbb{P}(A) \cap \mathbb{P}(B)$.
    e) $\mathbb{P}(A \cap B) \subseteq \mathbb{P}(A \cup B)$.

13. Let A, B, C, D be sets with $A \subset C$ and $B \subset D$. Prove or give a counterexample for the conjecture: $A \cap B \subset C \cap D$.

14. Let A, B be sets of propositions. We say that A *is stronger than* B, denoted by

$$A \Longrightarrow B,$$

if and only if

$$\forall p \in B, \exists q_1, q_2, \ldots, q_n \in A \ni (q_1 \wedge q_2 \wedge \cdots \wedge q_n) \Rightarrow p.$$

Thus, if A is stronger than B, then every proposition in B is a logical consequence of the conjunction of some of the propositions in A. For example, if

$$A = \{p \vee q, \neg q, r \rightarrow q\},$$
$$B = \{p, \neg r, \neg q, s \vee \neg q\},$$
$$C = \{p \vee q, q\},$$

then $A \Longrightarrow B$ but $A \not\Longrightarrow C$.
    a) State in symbols and English $\neg(A \Longrightarrow B)$.
    b) Give (another) example of sets of propositions A, B, C such that $A \Longrightarrow B$ but $A \not\Longrightarrow C$.
    c) Show that for any set of propositions A, $A \Longrightarrow \emptyset$.
    d) Show that for any set of propositions A, $A \Longrightarrow A$.
    e) Show that if A and B are any sets of propositions, $A \subseteq B$ implies $B \Longrightarrow A$.
    f) If $A \Longrightarrow B$ and $B \Longrightarrow A$ must it be the case that $A = B$?
    g) If $A \Longrightarrow B$ and $C \Longrightarrow D$ does $A \cup C \Longrightarrow B \cup D$?

2.2 TRUTH SETS

As an application of our recently learned set theory we will use sets to help us understand propositional functions and quantifiers a little better. First, a definition:

> **Definition 2.5:** Let p be a propositional function with domain D. The *truth set of p* is
>
> $$\{x \in D : p(x) \text{ is true}\}.$$

We will usually denote the truth set of a propositional function by using the corresponding uppercase letter; thus the truth set for p would be P and that for q, Q. Note that if p has domain D then $P \subseteq D$.

As an example, let $D = \{1, 2, 3, 4, 6\}$, $p(x)$ be "x is even," and $q(x)$ be "x is a prime." Then we have

$$P = \{2, 4, 6\},$$

$$Q = \{2, 3\}.$$

For another example, let $D = \mathbb{R}$ and $p(x)$ be "$x^2 - 3x + 2 = 0$" and $q(x)$ be "$\sin^2 x + \cos^2 x = 1$." Then the truth set for p is $\{1, 2\}$ while the truth set for q is $\mathbb{R}$. In algebra you may have called these "solution sets" and you might have noted that "identities" were just those equations whose solution sets were $\mathbb{R}$, as q above.

We can use our set operations to express the truth sets of compound propositional functions. It should be clear that if P, Q correspond (respectively) to the truth sets of propositional functions p, q then

$$P \cap Q = \{x : p(x) \wedge q(x)\} \text{ is the truth set for } p(x) \wedge q(x),$$

$$P \cup Q = \{x : p(x) \vee q(x)\} \text{ is the truth set for } p(x) \vee q(x) \text{ and}$$

$$P^C = \{x : \neg p(x)\} \text{ is the truth set for } \neg p(x).$$

What about the truth set for $p(x) \to q(x)$? If we recall that

$$(p \to q) \iff (\neg p \vee q)$$

then we see that

$$P^C \cup Q = \{x : \neg p(x) \vee q(x)\} \text{ is the truth set for } p(x) \to q(x).$$

As an example of this, let $D = \{1, 2, 3, 4, 5, 6\}$, $p(x)$ be "x is even," $q(x)$ be "x is odd" and $r(x)$ be "x is 2 or 3." Then (you should check these yourself):

The truth set of $p(x) \vee q(x)$ is $P \cup Q = D$,

The truth set of $p(x) \wedge q(x)$ is $P \cap Q = \varnothing$,

The truth set of $p(x) \to q(x)$ is $P^C \cup Q = \{1, 3, 5\}$ and

The truth set of $\neg r(x)$ is $R^C = \{1, 4, 5, 6\}$.

For another example from our algebraic past, let $D = \mathbb{R}$ and $p(x)$ be "$x^2 - 3x + 2 > 0$." We know from algebra that $p(x)$ is equivalent to "$(x - 2)(x - 1) > 0$." If we let $p_1(x)$ be "$x - 2 > 0$," $p_2(x)$ be "$x - 1 > 0$," $p_3(x)$ be "$x - 2 < 0$" and $p_4(x)$ be "$x - 1 < 0$," then $p(x)$ is equivalent to

$$[p_1(x) \text{ and } p_2(x)] \quad \text{or} \quad [p_3(x) \text{ and } p_4(x)]$$

(this because a product with two factors is positive if and only if both factors have the same sign). Since $P_1 = (2, \infty)$, $P_2 = (1, \infty)$, $P_3 = (-\infty, 2)$ and $P_4 = (-\infty, 1)$ (see exercise 3, section 2.1 for a review of interval notation) the truth set for $p(x)$ is

$$(P_1 \cap P_2) \cup (P_3 \cap P_4)$$

which is

$$[(2, \infty) \cap (1, \infty)] \cup [(-\infty, 2) \cap (-\infty, 1)] = (2, \infty) \cup (-\infty, 1) = \mathbb{R} - [1, 2].$$

It should be clear what quantifiers mean in terms of truth sets:

$$\forall x \in D, p(x) \leftrightarrow P = D \text{ and}$$

$$\exists x \in D \ni p(x) \leftrightarrow P \neq \varnothing.$$

Further, we can now be more explicit about what we meant above when we said that two propositional functions were equivalent; we meant that their truth sets were equal; that is,

$$[p(x) \leftrightarrow q(x)] \leftrightarrow P = Q.$$

We can also use truth sets to shed light on some of the equivalences and implications involving quantifiers from the last chapter. For example,

$$\forall x \in D, p(x) \wedge q(x)$$

will be true if and only if

$$P \cap Q = D.$$

But this is equivalent to

$$P = D \text{ and } Q = D,$$

so

$$[\forall x \in D, p(x)] \wedge [\forall x \in D, q(x)]$$

is also true and hence

$$[\forall x \in D, p(x) \wedge q(x)] \leftrightarrow [[\forall x \in D, p(x)] \wedge [\forall x \in D, q(x)]].$$

In a similar vein,

$$\exists x \in D \ni p(x) \vee q(x)$$

will be true if and only if

$$P \cup Q \neq \emptyset,$$

i.e., when

$$P \neq \emptyset \text{ or } Q \neq \emptyset;$$

hence, this is equivalent to

$$[\exists x \in D \ni p(x)] \vee [\exists x \in D \ni q(x)].$$

Further,

$$\forall x \in D, p(x) \rightarrow q(x) \text{ ("every } p \text{ is a } q\text{")}$$

will be true when

$$P^C \cup Q = D.$$

But this will be the case exactly when the shaded region in the Venn diagram below is empty; that is, when $P \subseteq Q$.

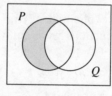

$$(P^C \cup Q)^C$$

However,

$$[\forall x \in D, p(x)] \rightarrow [\forall x \in D, q(x)]$$

will be true when $P = Q = D$ or when $P \neq D$ (Q can be anything in this case). $P \subseteq Q$ implies this condition so that the first proposition is stronger than the second (as we observed in section 1.7).

Exercises 2.2

1. Let $D = \{1, 2, 3, 4, 5, 6, 7, 8\}$, $p(x)$ be "x is even," $q(x)$ be "x is odd" and $r(x)$ be "x is a prime." Find:
   a) The truth set for $p(x) \wedge q(x)$.
   b) The truth set for $r(x) \rightarrow \neg p(x)$.
   c) $P^C \cup Q$.

d) A propositional function which has $\{1, 2, 3, 5, 7\}$ as its truth set.
e) The truth value of $\exists x \in D \ni r(x) \to p(x)$.
f) The truth value of $\forall x \in D, p(x) \lor q(x)$.
g) The truth value of $[\forall x \in D, p(x)] \lor [\forall x \in D, q(x)]$.
h) The truth value of $[\forall x \in D, q(x)] \to [\forall x \in D, r(x)]$.

2. Find the truth set for the propositional function "$x^2 - x - 2 \le 0$." Take the domain to be $\mathbb{R}$.

3. Consider the following pairs of propositions. For each proposition in a pair determine conditions on P, Q which guarantee that it is true and show that whenever the second (of each pair) is true, the first must also be true. Give an example to show that the first may be true and the second false.
   a) $\forall x \in D, p(x) \lor q(x); \forall x \in D, p(x) \lor \forall x \in D, q(x)$.
   b) $\exists x \in D \ni p(x) \land \exists x \in D \ni q(x); \exists x \in D \ni p(x) \land q(x)$.
   c) $\exists x \in D \ni p(x) \to q(x); \exists x \in D \ni p(x) \to \exists x \in D \ni q(x)$.

4. *Believe It or Not:* Conjecture: Let A, B be sets with $B \subseteq A$ and p a propositional function. Then $\forall x \in A, p(x)$ implies $\forall x \in B, p(x)$.

   "Proof": Suppose that A, B are as above and the given implication is false. Then there exists a $z \in B$ such that $p(z)$ is false. Since $B \subseteq A$, $z \in A$. But this means $\forall x \in A, p(x)$ is false, a contradiction. $\square$

   "Counterexample": Let $p(x)$ be "$x < 2$," $A = \{1, 2, 3\}, B = \emptyset$. Then $B \subseteq A, \forall x \in A, p(x)$ is false and $\forall x \in B, p(x)$ is true.

2.3 RELATIONS

We know that a set is determined by its elements; that is, $\{a, b\} = \{b, a\}$ and the order in which the elements are listed makes no difference. Sometimes, however, we wish to distinguish between the same elements listed in different order. To do this we introduce the concept of an *ordered pair*. While it is possible to define ordered pairs in terms of sets (see the exercises at the end of this section), this definition is not very useful so we will consider *ordered pair* to be an undefined term and instead define the main property we wish ordered pairs to have. Our notation is standard: the ordered pair with first element a and second element b will be denoted by (a, b). The property in which we are interested is

Definition 2.6: Let (a, b) and (c, d) be ordered pairs. Then $(a, b) = (c, d)$ if and only if $a = c$ and $b = d$.

We see that this indeed does distinguish order as $(a, b) \neq (b, a)$ unless $a = b$. With this concept in hand we can define a new operation on sets; the Cartesian product (sometimes referred to as *cross product* or simply *product*) of two sets:

Definition 2.7: Let A, B be sets. The *Cartesian product* of A with B, denoted by $A \times B$, is the set of all ordered pairs with first element in A and second element in B. In symbols,

$$A \times B = \{(a, b) : a \in A \text{ and } b \in B\}.$$

For example, if $A = \{1, 2, 3\}$, $B = \{a, b\}$ and $C = \varnothing$ then

$$A \times B = \{(1, a), (1, b), (2, a), (2, b), (3, a), (3, b)\},$$
$$B \times A = \{(a, 1), (a, 2), (a, 3), (b, 1), (b, 2), (b, 3)\},$$
$$A \times C = \varnothing,$$
$$B \times C = \varnothing.$$

We can illustrate $A \times B$ with a rectangular array:

$$
\begin{array}{c|ccc}
b & (1,b) & (2,b) & (3,b) \\
B \quad a & (1,a) & (2,a) & (3,a) \\
\hline
 & 1 & 2 & 3 \\
 & & A &
\end{array}
$$

Note that $A \times B \neq B \times A$ and that $A \times C = B \times C$ does not imply that $A = B$.

Quite often in ordinary English (as distinguished from the language mathematicians speak) we refer to objects being related to one another. For example, we might say, "She is related to me; she is my aunt" or "The grades I get are related to the amount of time I study." In mathematics, relations between objects are very important and we want to make this concept more precise. The following definition does this, although it probably is not immediately obvious that it embodies our usual idea of relation.

Definition 2.8: Let A, B be sets. A *relation from A to B* is a subset of $A \times B$. If R is a relation from A to B and $(a, b) \in R$ we will denote this by aRb. The *domain* of R (denoted by $Dom(R)$) is the set of all first elements of elements of R; in symbols

$$Dom(R) = \{a : (a, b) \in R\}$$
$$= \{a : aRb\}.$$

The *image* of R (denoted by $Im(R)$) is the set of all second elements of elements of R; in symbols

$$Im(R) = \{b : (a, b) \in R\}$$
$$= \{b : aRb\}.$$

Note that $Dom(R) \subseteq A$ and $Im(R) \subseteq B$. If $A = B$ we say that R is a *relation on A*.

As an example of a relation, suppose that $A = \{1, 2, 3\}$ and R is the "less than" relation on A; that is, aRb if and only if $a < b$. We can illustrate this with a diagram, the ○ representing elements of $A \times A$ and the ▢ squares indicating those ordered pairs in R:

In this example $Dom(R) = \{1, 2\}$ and $Im(R) = \{2, 3\}$.

Examples

Before we proceed any further, a few more examples may help us fix these new ideas in our minds.

1. Let A be the set of all living persons in the world and for $x, y \in A$ define xRy if and only if y is a parent of x. Then R is a relation on A. The ordered pairs in R are of the form $(x, x\text{'s parent})$. $Dom(R) = \{x : \text{one of } x\text{'s parents is alive}\}$ and $Im(R) = \{x : x \text{ is a parent with a living child}\}$.

2. Let $A = \mathbb{R}$ (the set of real numbers, a familiar friend from precalculus days) and for x, $y \in \mathbb{R}$ define xRy if and only if $y = x^2$. Then R is a relation on $\mathbb{R}$ and the ordered pairs in R are of the form (x, x^2). In fact, the ordered pairs in R are just the ordered pairs in the function $y = x^2$, the familiar parabola. Indeed, all the functions of precalculus and calculus are relations on $\mathbb{R}$. $Dom(R) = \mathbb{R}$ and $Im(R) = \{x : x \geq 0\}$.

3. Let A be any set and for x, $y \in A$ define xRy if and only if $x = y$. Then R is a relation on A. The ordered pairs in R are of the form (x, x). Here, $Dom(R) = Im(R) = A$.

4. Let A be any set. If B, C are subsets of A we say BRC if and only if $B \subseteq C$. Then R is a relation on $\mathbb{P}(A)$ and $Dom(R) = Im(R) = \mathbb{P}(A)$. In particular, if $A = \{1, 2\}$ then

$$R = \{(\varnothing, \varnothing), (\varnothing, \{1\}), (\varnothing, \{2\}), (\varnothing, A),$$
$$(\{1\}, \{1\}), (\{1\}, A), (\{2\}, \{2\}), (\{2\}, A), (A, A)\}.$$

5. Let A be the set of people in the USA and let B be the set of positive integers less than 10^{10}. Then for $x \in A$ and $y \in B$ we say xRy if and only if y is x's social security number. Then R is a relation from A to B. Ordered pairs in R are of the form (person, person's social security number). $Dom(R) = \{x : x$ has a social security number$\}$ and $Im(R) = \{x : x$ is someone's social security number$\}$. We won't list all the elements of R here; for further information see the Social Security Administration.

6. Let $A = \{1, 2, 3\}$ and let $B = \{1, 2\}$. Then $R = \{(3, 1), (3, 2)\}$, $S = \varnothing$, $T = A \times B$ are all relations from A to B. $Dom(R) = \{3\}$, $Im(R) = \{1, 2\}$, $Dom(S) = Im(S) = \varnothing$, $Dom(T) = A$, $Im(T) = B$. Note that relations need not "make sense" or have any special rule or pattern; the definition allows *any* subset of $A \times B$ to be a relation from A to B.

7. Let A, B be sets of propositions and for $p \in A$, $q \in B$ define pRq if and only if $p \rightarrow q$ is a tautology. Then R is a relation from A to B and an ordered pair $(p, q) \in R$ if and only if q is a logical consequence of p. We can think of $Im(R)$ as the set of all conclusions which can be logically implied from individual elements in A.

8. Let A be the set of all triangles in the plane (friends from high school geometry). If s, t are in A we will say sRt if and only if s is similar to t ($s \sim t$ in our high school notation). Then R is a relation on A with $Dom(R) = Im(R) = A$ (since every triangle is similar to itself).

9. Let $\mathbb{R}$ be the set of real numbers and for x, y in $\mathbb{R}$ define xRy if and only if $x \leq y$ (note: $x \leq y$ if and only if $x < y$ or $x = y$). Then R is a relation on $\mathbb{R}$ with $Dom(R) = Im(R) = \mathbb{R}$. Some typical elements in R are $(2, 3)$, $(2, 2)$, $(\pi, 10089\frac{1}{2})$.

10. Let $\mathbb{Z}$ be the set of integers and for x, y in $\mathbb{Z}$ we define xRy if and only if x divides y, denoted by $x \mid y$ (note: this is defined as: $x \mid y \leftrightarrow \exists z$ an integer $\ni y = xz$. Thus $3 \mid 6$, $2 \mid 8$, $-3 \mid 6$, $3 \mid -9$, $2 \mid 0$, while 2 does not divide 9—denoted by $2 \nmid 9$). Then R is a relation on $\mathbb{Z}$ with $Dom(R) = Im(R) = \mathbb{Z}$ (every integer divides itself). Some typical elements in R are $(1, 3)$, $(7, 21)$, $(1001, 1001)$, $(-1, 3)$.

11. Let $\mathbb{N}$ be the set of natural numbers and for x, y in $\mathbb{N}$ we define xRy if and only if $5 \mid (x - y)$. Then R is a relation on $\mathbb{N}$ with $Dom(R) = Im(R) = \mathbb{N}$. Some typical elements in R are $(10, 5)$, $(5, 10)$, $(3, 43)$, $(482, 257)$.

Observe that our way of writing relations, xRy, which might have seemed strange at first, is actually the way we usually write some well-known relations: $=$ (is equal to), $\leq$ (is less than or equal to), $\subseteq$ (is a subset of), $\sim$ (is similar to), $\mid$ (divides), $\Longleftrightarrow$ (is logically equivalent to). We should also note that many of the above examples are familiar to us, we just did not know that they were called relations!

There are certain properties which a relation on a set may or may not have; some of the more important ones are given names below:

Definition 2.9: Let R be a relation on a set A. Then we say:

a) R is *reflexive* if and only if $\forall a \in A$, aRa. $(a,a) \in R$

b) R is *symmetric* if and only if $\forall a, b \in A$, $aRb \rightarrow bRa$.

c) R is *transitive* if and only if $\forall a, b, c \in A$, $(aRb \wedge bRc) \rightarrow aRc$.

d) R is *antisymmetric* if and only if $\forall a, b \in A$, $(aRb \wedge bRa) \rightarrow a = b$.

e) R is *irreflexive* if and only if $\forall a \in A$, $\neg(aRa)$. no $(a,a) \in R$

f) R is *complete* if and only if $\forall a, b \in A$, $a \neq b \rightarrow (aRb \vee bRa)$.

g) R is *asymmetric* if and only if $\forall a, b \in A$, $aRb \rightarrow \neg(bRa)$.

h) R is an *equivalence relation* if and only if R is reflexive, symmetric and transitive.

i) R is a *partial order* if and only if R is reflexive, transitive and antisymmetric.

j) R is a *strict partial order* if and only if R is irreflexive and transitive.

k) R is a *total order* (or *linear order*) if and only if R is a partial order which is complete.

l) R is a *strict total order* if and only if R is a strict partial order which is complete.

It may be helpful to try to characterize some of these properties in an informal manner so that they do not seem so strange. If R is reflexive then everything in A is related to itself. If R is symmetric then whenever a is related to b we must have b related to a. A common example of a transitive relation is "preference"; i.e., if I prefer apple pie to chocolate cake (a difficult decision!) and I prefer chocolate cake to wilted lettuce (an easy decision) then I should prefer apple pie to wilted lettuce. If R is an antisymmetric relation then the only way we can have both aRb and bRa is to have $a = b$. An example of this is the relation $\subseteq$; in fact, this is the very property which we use to prove set equality. R is irreflexive if no element of A is related to itself. The "parent" relation is irreflexive. R is complete if given any two distinct elements in A the first must be related to the second or vice-versa. An example of a relation which is *not* complete is $\subseteq$ on the power set of a set with at least two elements (try to find two subsets which are not related). We normally think of preference as being an asymmetric relation; if I prefer apple pie to chocolate cake then I don't also prefer chocolate cake to apple pie. Please note that symmetric and antisymmetric are not mutually exclusive; a relation may have both of these properties (or neither of them, for that matter).

This might be a good time to make some general remarks about mathematical definitions. It is often the case (as in the above examples) that definitions give names to objects which have certain properties. This means that if an object has the defining property for a "whatever," then we will call it "whatever," and if we call an object a "whatever," then it has the defining property of a "whatever." That is why we used "if and only if." If you look through many mathematics books (and on occasion, this book too—be on the lookout!) you will find definitions given using just "if—then," rather than "if and only if." This is a convention of mathematical writing and should be understood as "if and only if," even if the "only if" doesn't appear explicitly.

There is another feature of mathematical definitions which many students find disconcerting. When we go to a dictionary to look up the definition of a new word, we expect to come away with some idea of what the word means. On the other hand, a mathematical definition very seldom gives much help in the way of meaning; it usually states some cryptic defining properties and the actual consequences of these properties (and hence the "meaning") are developed by the theorems and examples which follow. If we think back to the definition of *derivative* in calculus, there was not much meaning in that limit; in fact, most of the effort in a first course in calculus is directed towards giving meaning to this definition. Are mathematical definitions important, even if they don't immediately contribute to our understanding? Yes, they are very important. Without precise definitions we cannot carry out any proofs; thus, we should not expect to be able to prove that a given relation is reflexive unless we know the definition of *reflexive*. Should we expect to get much meaning out of definitions? No, that is not their purpose. The meaning of mathematical terms is developed through theorems and examples.

When confronted with the definition of a new concept, a useful activity is to try to construct examples of appropriate objects which satisfy the definition and examples which do not. This helps make some of the consequences of the definition more apparent and helps to make the concept more concrete. To help establish this good habit, let's start now by looking at a few examples related to the definitions above (you should verify the assertions made and make up some more examples of your own): with $A = \{1, 2, 3, 4\}$, let

$$R = \{(1, 2), (2, 3)\},$$

$$S = \{(1, 1), (2, 2), (1, 2), (2, 1), (3, 4)\},$$

$$T = \{(1, 1), (2, 2), (3, 3), (4, 4)\}.$$

Then: R is not reflexive, not symmetric, not transitive, antisymmetric, irreflexive, not complete, asymmetric; S is not reflexive, not symmetric, transitive, not antisymmetric, not irreflexive, not complete, not asymmetric; T is reflexive, symmetric, transitive, antisymmetric, not irreflexive, not complete, not asymmetric.

It may be helpful to have some pictorial views of these properties. With $A = \{1, 2, 3, 4\}$, then if R is reflexive, it must contain at *least* the main diagonal (the squares in the figure below):

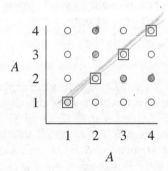

If R is symmetric, then its graph must be symmetric with respect to the main diagonal; i.e., if $(2, 3)$ and $(4, 2)$ are elements of R then $(3, 2)$ and $(2, 4)$ must be in R also:

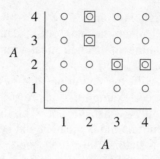

We see that the examples given earlier also satisfy some of these properties; for example, $=$ and $\sim$ are equivalence relations, $\leq$ and $\subseteq$ are partial orders, $<$ and $\subset$ are strict partial orders, $\leq$ is a total order and $<$ is a strict total order. In fact, we can think of "equivalence relation" as abstracting the idea of equality and "partial order" as abstracting the idea of ordering we have on the real numbers. This gives a good example of the power of abstraction; we can think of equivalence relations as containing the "essence" of the idea of equality (reflexive, symmetric and transitive) and not be distracted by any particular features a certain embodiment of it may have. As you might imagine, later on we will prove some interesting and useful facts about equivalence relations. In the meantime we have some exercises to do to check our understanding of this material. But first, an example of a theorem and proof involving some of these ideas and a discussion of proof forms.

Theorem 2.5: Let A be a non-empty set. Suppose that R is the relation on $\mathbb{P}(A)$ defined by BRC if and only if $B \subset C$. Then R is transitive and irreflexive; i.e., $\subset$ is a strict partial order.

Proof: Suppose that B, C and D are subsets of A with $B \subset C$ and $C \subset D$. To show that $\subset$ is transitive we must show that $B \subset D$. Let $a \in B$. Then since $B \subset C$, $a \in C$. Also, since $C \subset D$, $a \in C \rightarrow a \in D$. We are nearly done, as we now have $B \subseteq D$ and all that remains is to show that B is a proper subset of D; that is, $B \neq D$. But since $B \subset C$ there exists an $x \in C$ such that $x \notin B$. However, x must also be an element of D as $C \subset D$. Thus $B \neq D$ and $\subset$ is transitive. To show that $\subset$ is irreflexive, let $B \in \mathbb{P}(A)$. Then $B \not\subset B$ (since $B = B$) so we have $\forall B \in \mathbb{P}(A),\ \neg(B \subset B)$. $\square$

Let's pause for a moment to think about the forms for proofs of these properties. We will consider some representative types, leaving some of the others as exercises. Suppose that R is a relation on a non-empty set A. A direct proof that R is reflexive would have the form: Let $a \in A$.

$$\cdots$$

"something or other depending on R"

$$\cdots$$

so $(a, a) \in R$ and R is reflexive.

For a proof of symmetry: Let $a, b \in A$ with $(a, b) \in R$.

$$\cdots$$

"something or other depending on R"

$$\cdots$$

so $(b, a) \in R$ and R is symmetric.

For transitivity: Let $a, b, c \in A$ with (a, b) and $(b, c) \in R$.

$$\cdots$$

"something or other depending on R"

$$\cdots$$

so $(a, c) \in R$ and R is transitive.

For completeness: Let $a, b \in A$ with $a \neq b$. Suppose $(a, b) \notin R$.

$$\cdots$$

"something or other depending on R"

$$\cdots$$

so $(b, a) \in R$ and R is complete.

Note in this last proof we used our usual technique for a conclusion which is a disjunction; we assumed that one subproposition was false and then proved that the other must hold. It is also worth noting that in each case our starting point for the proof was determined by the conclusion and not the hypotheses, which would have to do with whatever properties R had.

Exercises 2.3

1. Let $A = \{a, b, c\}$, $B = \{1, 2\}$, $C = \{4, 5, 6\}$.
   a) List the elements of $A \times B$, $B \times A$, $A \times C$.
   b) Give examples of relations from A to B and B to A, each of which have four elements.
   c) Give an example of a symmetric relation on C which has three elements.

2. Suppose $A = \{1, 2, 3\}$. For each of the relations R given below, list the elements of R, find $Dom(R)$ and $Im(R)$ and tell which of the properties of definition 2.9 R has.
   a) R is the relation $<$ on A.
   b) R is the relation $\geq$ on A.
   c) R is the relation $\subset$ on $\mathbb{P}(A)$.

3. Let A, B, C, D be sets. Prove or give counterexamples for the following conjectures:
   a) $A \times (B \cup C) = (A \times B) \cup (A \times C)$.
   b) $A \times (B \cap C) = (A \times B) \cap (A \times C)$.
   c) $(A \times B) \cap (A^C \times B) = \emptyset$.
   d) $(A \subseteq B \wedge C \subseteq D) \rightarrow A \times C \subseteq B \times D$.
   e) $A \cup (B \times C) = (A \cup B) \times (A \cup C)$.
   f) $A \cap (B \times C) = (A \cap B) \times (A \cap C)$.
   g) $(A \times B) \cap (C \times D) = (A \cap C) \times (B \cap D)$.
   h) $A \times (B - C) = A \times B - A \times C$.

4. Suppose that R is a relation on a non-empty set A. Give the form of a direct proof that:
   a) R is antisymmetric.
   b) R is irreflexive.
   c) R is asymmetric.

5. Let A, B be sets with R, S relations from A to B. Prove:
   a) $Dom(R \cup S) = Dom(R) \cup Dom(S)$.
   b) $Dom(R \cap S) \subseteq Dom(R) \cap Dom(S)$ and give an example to show that equality need not hold.
   c) $Im(R \cup S) = Im(R) \cup Im(S)$.
   d) $Im(R \cap S) \subseteq Im(R) \cap Im(S)$ and give an example to show that equality need not hold.

6. Let A be a non-empty set. Show that:
   a) If $R = A \times A$ then R is reflexive, symmetric, transitive and complete. What can be said about whether R is asymmetric or antisymmetric?
   b) If $R = \emptyset$ then R is symmetric, transitive, asymmetric, antisymmetric, irreflexive but not reflexive.
   c) If $R = \{(a, a) : a \in A\}$ then R is an equivalence relation and is also antisymmetric but not asymmetric.

7. Referring to the examples given earlier in this section, show that:
   a) Example 1 is asymmetric but is not reflexive.
   b) Examples 3, 11 are equivalence relations.
   c) Examples 3, 4, 9 are partial orders.
   d) Examples 2, 10 are not complete.
   e) Example 9 is a total order.

8. Let R be the equivalence relation given in example 11 above. Determine all the elements in these sets:

a) $\{x : xR1\}$.

b) $\{x : xR2\}$.

c) $\{x : xR7\}$.

9. Let R be the relation $\mid$ on $\mathbb{Z}$ described in example 10 above.

a) List three elements of $\mathbb{Z} \times \mathbb{Z}$ which are not elements of R.

b) Which of $(0, 0), (0, 1), (1, 0)$ are elements of R?

c) Prove the following:

i) $\forall n \in \mathbb{Z}, n \mid 0$.

ii) $\forall n \in \mathbb{Z}, 0 \mid n \to n = 0$.

iii) $\forall a, b, c \in \mathbb{Z}, (a \mid b \wedge a \mid c) \to a \mid (b + c)$.

iv) $\forall a, b, c \in \mathbb{Z}, a \mid b \to a \mid bc$.

10. Let R, S be relations on a non-empty set A. Prove or give counterexamples for the following conjectures:

a) R is complete $\to R$ is reflexive. $(1,2)\ (2,1)$ *example counter*

b) R is transitive and irreflexive $\to R$ is asymmetric.

c) R is reflexive $\to R$ is not asymmetric.

d) R is asymmetric $\to R$ is not reflexive.

e) $Dom(R) \cap Im(R) = \varnothing \to R$ is transitive, antisymmetric, irreflexive and asymmetric.

f) R a strict partial order $\to R$ is antisymmetric and asymmetric.

g) R not reflexive $\to R$ is irreflexive.

h) R and S symmetric $\to R \cap S$ symmetric.

i) R or S symmetric $\to R \cap S$ symmetric.

j) R and S symmetric $\to R \cup S$ symmetric.

k) R or S reflexive $\to R \cup S$ reflexive.

l) R and S transitive $\to R \cup S$ transitive.

m) R and S transitive $\to R \cap S$ transitive.

11. Give examples (if possible) of relations R which are

a) Reflexive and symmetric but not transitive.

b) Symmetric and transitive but not reflexive.

c) Asymmetric but not antisymmetric.

d) Symmetric and antisymmetric.

e) Neither reflexive nor irreflexive.

12. If R is a relation from A to B and $C \subseteq A$ we define the *restriction of R to C*, denoted by $R|_C$, as

$$\{(x, y) \in R : x \in C\}.$$

a) Let $A = B = \{1, 2, 3, 4\}, C = \{2, 4\}$. Let R be the relation $<$ from A to B. Find R and $R|_C$.

b) If R is a relation from A to B and $C \subseteq A$ show that $Dom(R|_C) = Dom(R) \cap C$.

c) If R is a relation on A and $B \subseteq A$, is $R|_B$ a relation on B? Prove or give a counterexample.

13. (Continuation of exercise 12.) Suppose that R is a relation on A with the properties listed below. If $B \subseteq A$ and $R|_B$ is considered as a relation on A, which of these properties must $R|_B$ also have? Prove or give counterexamples.
    a) Symmetric.
    b) Transitive.
    c) Antisymmetric.

14. Suppose that we define the ordered pair (a, b) by

$$(a, b) = \{\{a\}, \{a, b\}\}.$$

Show that with this definition we have

$$(a, b) = (c, d) \leftrightarrow (\acute{a} = c \wedge b = d).$$

15. Suppose that we define "ordered triple" using ordered pairs as

$$(a, b, c) = ((a, b), c).$$

Show that this has the desired ordered property; i.e.,

$$(a, b, c) = (d, e, f) \text{ if and only if } a = d, b = e, c = f.$$

16. Suppose that R is a strict total order on a non-empty set A. Show that R has the "trichotomy" property; i.e.,

$$\forall a, b \in A, \text{ exactly one of the following is true :}$$

$$a = b, \ aRb, \ bRa.$$

17. Let R be a relation on a non-empty set A. The *transitive closure* of R is the smallest transitive relation containing R; i.e., if S is the transitive closure of R, and T is any transitive relation containing R, then

$$R \subseteq S \subseteq T.$$

We make similar definitions for the reflexive and symmetric closures. We will denote these closures by R_{trans}, R_{sym} and R_{ref}.
    a) If $A = \{1, 2, 3, 4\}$ and $R = \{(1, 2), (1, 4), (2, 3)\}$ find R_{trans}, R_{sym} and R_{ref}.
    b) Prove or give counterexamples for the following conjectures (R, S are relations on a non-empty set A):
       i) $(R \cup S)_{trans} = R_{trans} \cup S_{trans}$.
       ii) $(R \cap S)_{trans} = R_{trans} \cap S_{trans}$.
       iii) $(R \cup S)_{sym} = R_{sym} \cup S_{sym}$.

iv) $(R \cap S)_{sym} = R_{sym} \cap S_{sym}$.

v) $(R \cup S)_{ref} = R_{ref} \cup S_{ref}$.

vi) $(R \cap S)_{ref} = R_{ref} \cap S_{ref}$.

c) What can be said about the corresponding concepts of antisymmetric and asymmetric closure?

18. Let R be a relation on a non-empty set A. Suppose that R is asymmetric and also satisfies the condition (sometimes called *negative transitivity*):

$$\forall x, y, z \in A, xRz \rightarrow (xRy \lor yRz).$$

a) Show that R is transitive. Such relations are sometimes called *weak orders*.

b) If R is transitive and asymmetric must it also satisfy the condition given above? Prove or give a counterexample.

19. Let R be a relation on a non-empty set A. Let $x \in A$. We define the R-class of x, denoted by $<x>_R$, as

$$<x>_R; = \{y : yRx\}.$$

a) Let $A = \{1, 2, 3, 4\}$ and

$$R = \{(1, 2), (1, 3), (2, 1), (1, 1), (2, 3), (4, 2)\}.$$

Find $<1>_R$, $<2>_R$, $<3>_R$ and $<4>_R$.

b) Show that R is reflexive iff $\forall x \in A, x \in <x>_R$.

c) Show that R is symmetric iff $\forall x, y \in A, x \in <y>_R \rightarrow y \in <x>_R$.

d) Show that $\forall x \in A, <x>_R \neq \emptyset$ iff $Im(R) = A$.

e) Suppose that $Dom(R) = A$ and R is symmetric and transitive. Show that

$$\forall x, y \in A, <x>_R \subseteq <y>_R \rightarrow xRy.$$

Also show

$$<x>_R \subseteq <y>_R \rightarrow <x>_R = <y>_R.$$

f) Suppose that R is symmetric and transitive. Show that

$$\forall x, y \in A, <x>_R \cap <y>_R \neq \emptyset \rightarrow <x>_R = <y>_R.$$

20. **Believe It or Not:** Conjecture: Suppose A and B are sets with $A \times B = B \times A$. Then $A = B$.

"Proof": Suppose $A \times B = B \times A$. Let $a \in A$, with $b \in B$ such that $(a, b) \in A \times B$. Since $A \times B = B \times A$, $(a, b) \in B \times A$. Thus $a \in B$ and $A \subseteq B$. A similar argument shows $B \subseteq A$. $\qquad\square$

"Counterexample": Let $A = \{1, 2, 3\}, B = \{\emptyset\}$. Then $A \times B = B \times A = \emptyset$ but $A \neq B$.

21. **Believe It or Not:** Conjecture: Suppose that R is a relation on a non-empty set A. If R is not symmetric then R is asymmetric.

"Proof": Let R be a relation on a non-empty set A. Suppose $a, b \in A$ with $(a, b) \in R$. Since R is not symmetric, $(b, a) \notin R$ so R is asymmetric.   □

"Counterexample": Let $A = \{1, 2, 3\}, R = \{(1, 2), (2, 1), (1, 3)\}$. Then R is neither symmetric nor asymmetric.

22. **Believe It or Not:** Conjecture: Let R be a relation on a non-empty set A. If R is symmetric and transitive then R is reflexive.

"Proof": Suppose that R is a symmetric, transitive relation on a non-empty set A. Let $a, b \in A$ with $(a, b) \in R$. Since R is symmetric, $(b, a) \in R$. But R is also transitive, so we have $(a, a) \in R$ and hence R is reflexive.   □

"Counterexample": Let $A = \{a, b, c\}, R = \{(a, b), (b, a), (a, c), (b, c), (a, a)\}$. Then R is symmetric and transitive but not reflexive since $(b, b) \notin R$.

2.4   MORE RELATIONS

Never satisfied with what we already have, we will continue in the tradition established for both propositions and sets and make new relations from old. Naturally we need some definitions:

Definition 2.10: Let R be a relation from A to B. R *inverse*, denoted by R^{-1}, is the relation from B to A given by $xR^{-1}y$ if and only if yRx. In symbols,

$$R^{-1} = \{(x, y) : (y, x) \in R\}.$$

We observe that $Dom(R^{-1}) = Im(R)$ and $Im(R^{-1}) = Dom(R)$.

For example, if R is the "parent" relation (example 1 in the previous section, $xRy \leftrightarrow y$ is a parent of x) then R^{-1} is the "child" relation; $xR^{-1}y$ if and only if y is a child of x. Thus, if $(x, x\text{'s father}) \in R$ then $(x\text{'s father}, x) \in R^{-1}$. As another example of this, if R is the relation on $\mathbb{N}$ given by xRy if and only if $x < y$ then $xR^{-1}y$ if and only if $y < x$.

Although we can use our set operations to get new relations from old (since relations are sets) we also have an operation peculiar to relations called composition:

> **Definition 2.11:** Let R be a relation from A to B and let S be a relation from B to C. Then R *composed with* S (denoted by $S \circ R$) is the relation from A to C given by
>
> $$S \circ R = \{(x, z) : \exists y \in B \ni [(x, y) \in R \wedge (y, z) \in S]\}.$$

The reason for the apparently reversed order of S and R in the above notation is due to the way we will write functions later on (which turns out to be the way you have written functions too: $f(x)$). Also observe that indeed $S \circ R$ is a relation from A to C for if $(x, y) \in R$ then $x \in A$ and if $(y, z) \in S$ then $z \in C$ so $S \circ R \subseteq A \times C$.

As an example of how one "builds" $S \circ R$ from R and S, let $A = \{1, 2, 3, 4\}$, $B = \{a, b, c\}$, $C = \{4, 5, 6\}$ and $R = \{(1, a), (1, b), (3, a)\}$, $S = \{(a, 5), (b, 4), (a, 6), (c, 6)\}$. Then we can think: "Which second coordinates of elements of R coincide with first coordinates of elements of S?" These will combine to produce elements of $S \circ R$. Thus $(1, a) \in R$ and $(a, 5) \in S$ gives us $(1, 5) \in S \circ R$. Continuing, we get

$$(1, 4) \in S \circ R \text{ from } (1, b) \in R \text{ and } (b, 4) \in S,$$

$$(3, 5) \in S \circ R \text{ from } (3, a) \in R \text{ and } (a, 5) \in S,$$

$$(1, 6) \in S \circ R \text{ from } (1, a) \in R \text{ and } (a, 6) \in S,$$

$$(3, 6) \in S \circ R \text{ from } (3, a) \in R \text{ and } (a, 6) \in S.$$

Thus $S \circ R = \{(1, 5), (1, 4), (3, 5), (1, 6), (3, 6)\}$.

One last definition (for a little while anyway) and we will be ready for some more examples:

> **Definition 2.12:** Let A be a set. The *identity* relation on A, denoted by I_A, is given by
>
> $$I_A = \{(x, x) : x \in A\}.$$

We should recognize this as our old relational friend "equality" clothed in a new coat of respectability. Thus $a I_A b$ if and only if $a = b$.

Example: Let $A = \{1, 2, 3\}$ and let R be the relation on A such that $R = \{(1, 2), (1, 3), (2, 3)\}$. Then:

a) $R^{-1} = \{(2, 1), (3, 1), (3, 2)\}$.

b) $I_A = \{(1, 1), (2, 2), (3, 3)\}$.

c) $R^{-1} \circ R = \{(1, 1), (1, 2), (2, 2), (2, 1)\}$.

d) $R \circ R^{-1} = \{(2, 2), (2, 3), (3, 3), (3, 2)\}$.

e) $R \circ R = \{(1, 3)\}$.

f) $R^{-1} \circ R^{-1} = \{(3, 1)\}$.

g) $R \circ I_A = I_A \circ R = \{(1, 2), (1, 3), (2, 3)\}$.

h) $R^{-1} \circ I_A = I_A \circ R^{-1} = \{(2, 1), (3, 1), (3, 2)\}$.

Some observations we can make about this example are: $R \circ R^{-1} \neq R^{-1} \circ R$ so that composition of relations is not commutative; it also appears that $R \circ I_A = I_A \circ R = R$ may be true for any relation R on a set A.

Another example: Let $A = \{1, 2, 3\}$, $B = \{4, 5, 6\}$, $C = \{2, 3, 4\}$ with

$$R = \{(1, 4), (1, 5), (2, 6), (3, 4)\}$$

a relation from A to B and

$$S = \{(4, 2), (4, 3), (6, 2)\}$$

a relation from B to C.

It may help to illustrate this with a diagram:

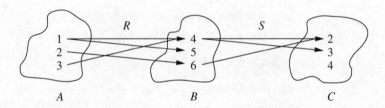

Some calculations reveal:

a) $S \circ R = \{(1, 2), (1, 3), (3, 2), (3, 3), (2, 2)\}$.

b) $R^{-1} = \{(4, 1), (5, 1), (6, 2), (4, 3)\}$.

c) $S^{-1} = \{(2, 4), (3, 4), (2, 6)\}$.

d) $R^{-1} \circ S^{-1} = \{(2, 1), (3, 1), (2, 3), (2, 2), (3, 3)\}$.

e) $(S \circ R)^{-1} = \{(2, 1), (3, 1), (2, 3), (3, 3), (2, 2)\}$.

We note that $R \circ S$ and $S^{-1} \circ R^{-1}$ are not defined and that $R^{-1} \circ S^{-1} = (S \circ R)^{-1}$.

For a little practice in theorem proving we will show that this last equality always holds:

Theorem 2.6: Let A, B, C be sets, R a relation from A to B and S a relation from B to C. Then $(S \circ R)^{-1} = R^{-1} \circ S^{-1}$.

Proof: It may be helpful to refer to the figure below to keep in mind the various sets and relations involved.

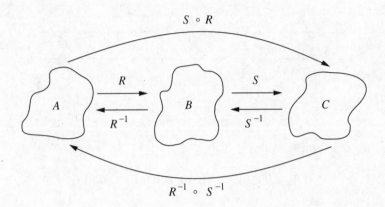

First, we observe that $(S \circ R)^{-1}$ is a relation from C to A as is $R^{-1} \circ S^{-1}$, so at least they have a chance of being equal. Recall that relations are sets, so to show that two relations are equal we must show that they are equal as sets. To start, let $(x, z) \in (S \circ R)^{-1}$. Then $(z, x) \in S \circ R$ so there exists a $y \in B$ such that $(z, y) \in R$ and $(y, x) \in S$. Hence $(y, z) \in R^{-1}$ and $(x, y) \in S^{-1}$. Therefore $(x, z) \in R^{-1} \circ S^{-1}$ so $(S \circ R)^{-1} \subseteq R^{-1} \circ S^{-1}$.

Now to show the set inclusion the other way round, let $(x, z) \in R^{-1} \circ S^{-1}$. Then there exists a $y \in B$ such that $(x, y) \in S^{-1}$ and $(y, z) \in R^{-1}$. Hence $(y, x) \in S$ and $(z, y) \in R$, so we have $(z, x) \in S \circ R$ and $(x, z) \in (S \circ R)^{-1}$ as desired. $\square$

Although the previous proof may look a little intricate, each step just involved translation of set membership using a definition; e.g., $(x, y) \in S^{-1}$ means $(y, x) \in S$. In words, this result tells us that the inverse of a composition of relations is the composition of the inverses in the opposite order.

Here is another example involving composition. Let $A = \{1, 2, 3\}$, $B = \{4, 5, 6\}$, $C = \{6, 7, 8\}$ and $D = \{1, 4, 6\}$ with

$$R = \{(1, 4), (3, 5), (3, 6)\} \text{ a relation from } A \text{ to } B,$$

$$S = \{(4, 6), (6, 8)\} \text{ a relation from } B \text{ to } C,$$

$$T = \{(6, 1), (8, 6), (6, 4)\} \text{ a relation from } C \text{ to } D.$$

These are illustrated in the figure below:

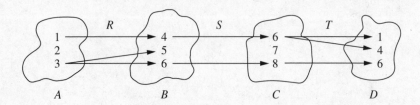

Then we can form

$$S \circ R = \{(1, 6), (3, 8)\} \text{ a relation from } A \text{ to } C$$

and

$$T \circ S = \{(4, 1), (4, 4), (6, 6)\} \text{ a relation from } B \text{ to } D.$$

Now these can be composed with T and R to obtain

$$T \circ (S \circ R) = \{(1, 1), (1, 4), (3, 6)\} \text{ a relation from } A \text{ to } D$$

and

$$(T \circ S) \circ R = \{(1, 1), (1, 4), (3, 6)\} \text{ a relation from } A \text{ to } D.$$

Amazingly, these two are equal! We were not exceptionally fortunate in our choices for R, S and T, however; this equality always holds. Mathematicians express this by saying "composition of relations is associative." Here is a proof of it:

Theorem 2.7: Let A, B, C, D be sets with R a relation from A to B, S a relation from B to C and T a relation from C to D. Then

$$T \circ (S \circ R) = (T \circ S) \circ R.$$

Proof: The figure below may help us to remember the various relations involved:

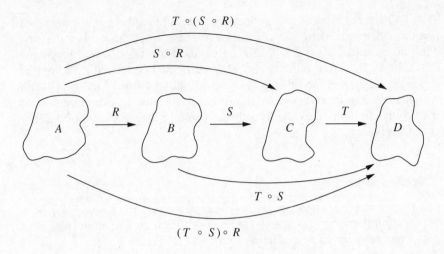

We need to show set inclusion both ways. First to see that $T \circ (S \circ R) \subseteq (T \circ S) \circ R$, let $(x, y) \in T \circ (S \circ R)$. Then $x \in A$ and $y \in D$ and there exists $z \in C$ such that $(x, z) \in S \circ R$ and $(z, y) \in T$. Since $(x, z) \in S \circ R$ there exists a $w \in B$ such that $(x, w) \in R$ and $(w, z) \in S$. Now $(w, z) \in S$ and $(z, y) \in T$ implies that $(w, y) \in T \circ S$. But $(x, w) \in R$ so $(x, y) \in (T \circ S) \circ R$. That the inclusion holds the other way round is left as an exercise. $\qquad \square$

We can also combine some of the special properties of relations with these operations. For example, we have

---

Theorem 2.8: Let R be a relation on A. Then R is transitive if and only if $R \circ R \subseteq R$.

Proof: As this is an equivalence, we have two implications to prove. We will start by showing that R transitive implies $R \circ R \subseteq R$. Let $(x, y) \in R \circ R$. Then there exists $z \in A$ such that $(x, z) \in R$ and $(z, y) \in R$. But since R is transitive this means that $(x, y) \in R$. Therefore, $R \circ R \subseteq R$.

Now to show that $R \circ R \subseteq R$ implies that R is transitive. Suppose that (x, y) and (y, z) are elements of R. Then $(x, z) \in R \circ R$ and since $R \circ R \subseteq R$, (x, z) must also be an element of R and hence R is transitive. $\qquad \square$

One cannot refrain from noticing once again that the form of the proof here was determined by the conclusion of each implication. For the first one, the conclusion was $R \circ R \subseteq R$ so we used our usual subset technique of starting with a fixed-but-arbitrary element in the one set and showing that it was an element of the other set. In the second implication the conclusion was that R is transitive so we showed that R satisfied the definition of transitivity. In both cases the hypotheses came in during the proof and not at the beginning.

Exercises 2.4

1. Let $A = \{1, 2, 4\}$, $B = \{1, 3, 4\}$. Let $R = \{(1, 3), (1, 4), (4, 4)\}$ be a relation from A to B, $S = \{(1, 1), (3, 4), (3, 2)\}$ be a relation from B to A and let $T = \varnothing$ be a relation from A to B. Find:
   a) $Dom(R)$.
   b) $Dom(S)$.
   c) $Dom(T)$.
   d) $Im(R)$.
   e) $Im(S)$.
   f) $Im(T)$.
   g) $S \circ R$.
   h) $R \circ S$.
   i) $Dom(S \circ R)$.
   j) $Im(S \circ R)$.
   k) $Dom(R \circ S)$.
   l) $Im(R \circ S)$.
   m) R^{-1}.
   n) S^{-1}.
   o) I_A.
   p) I_B.
   q) $R^{-1} \circ S^{-1}$.
   r) $S^{-1} \circ R^{-1}$.
   s) $(R \circ S)^{-1}$.
   t) $(S \circ R)^{-1}$.
   u) T^{-1}.
   v) I_B^{-1}.
   w) $(R \circ S) \circ R$.
   x) $R \circ (S \circ R)$.

2. Let R be a relation on a non-empty set A. Show that:
   a) $(R^{-1})^{-1} = R$.
   b) $I_A^{-1} = I_A$
   c) R is reflexive if and only if $I_A \subseteq R \subseteq R \circ R$.

d) R is symmetric if and only if $R = R^{-1}$.

e) R is transitive if and only if R^{-1} is transitive.

f) R is an equivalence relation if and only if R^{-1} is an equivalence relation.

g) Suppose that $Dom(R) = A$. R is an equivalence relation if and only if $R = R^{-1} = R \circ R$.

h) R is asymmetric if and only if $R \cap R^{-1} = \varnothing$.

i) $R \cup R^{-1} = A \times A$ implies R is complete.

j) R symmetric implies $R \circ R$ is symmetric.

k) $I_{Dom(R)} \subseteq R^{-1} \circ R$.

l) R is a partial order if and only if R^{-1} is a partial order.

m) R is a partial order if and only if $R \cap R^{-1} = I_A$ and $R \circ R = R$.

n) R is a strict partial order if and only if R^{-1} is a strict partial order.

3. Let A be a non-empty set with R, S relations on A. Consider the following conjectures. Prove the true ones and give counterexamples for those that are false.

a) R symmetric implies $R \circ R$ symmetric.

b) $R \circ S^{-1} = S \circ R^{-1}$ implies $R \circ S^{-1}$ is symmetric.

4. Let R be a relation from A to B and S a relation from B to C. Show that:

a) $Dom(S \circ R) \subseteq Dom(R)$.

b) $Im(S \circ R) \subseteq Im(S)$.

c) $Im(R) \subseteq Dom(S)$ implies $Dom(S \circ R) = Dom(R)$. Does the converse of this hold?

5. Complete the proof of theorem 2.7.

6. Suppose R and S are equivalence relations on a non-empty set A. Consider the following conjectures. Prove the true ones and give counterexamples for those that are false.

a) $R \cup S$ is an equivalence relation implies $R \circ S = S \circ R$.

b) $R \cup S = R \circ S$ implies $R \cup S$ is an equivalence relation.

c) $R \cup S = R \circ S$ implies $R \circ S = S \circ R$.

7. Let R be the relation $<$ on the integers. Show that R is a strict partial order. Also show that $R \cup I_{\mathbb{Z}}$ (which is $\leq$) is a partial order.

8. Let R be a partial order on a non-empty set A. Show that $R - I_A$ is a strict partial order on A.

9. Let R be a relation on a non-empty set A. Prove or give counterexamples (refer to exercise 17, section 2.3):

a) $R_{ref} = R \cup I_A$.

b) $R_{sym} = R \cup R^{-1}$.

c) $R_{trans} = R \cup (R \circ R)$.

10. Let R, S, T be relations between sets. Determine some conditions on R, S, T which will guarantee the following conclusions. Prove that your conjectures are correct.
    a) $R \circ S = R \circ T$ implies $S = T$.
    b) $S \circ R = T \circ R$ implies $S = T$.

11. **Believe It or Not:** Conjecture: Let R be a relation on a non-empty set A. If R is transitive then $R \circ R$ is transitive.

    "Proof": Let R be a transitive relation on A. Let $a, b, c \in A$ with $(a, b), (b, c) \in R \circ R$. Then there exist $d, e \in A$ such that $(a, d), (d, b), (b, e), (e, c) \in R$. Since R is transitive, $(a, b), (b, c) \in R$, which implies $(a, c) \in R \circ R$ so $R \circ R$ is transitive. $\qquad\square$

    "Counterexample": Let $A = \{1, 2, 3\}, R = \{(1, 2), (2, 2), (2, 3), (1, 3)\}$. Thus

    $$R \circ R = \{(1, 3), (1, 2), (2, 3), (2, 2)\},$$

    so while we have R transitive, $R \circ R$ is not.

12. **Believe It or Not:** Conjecture: Let R, S be equivalence relations on a non-empty set A. If $R \circ S = S \circ R$ then $R \cup S$ is an equivalence relation.

    "Proof": Let R, S be as above. Clearly, $R \cup S$ is reflexive. If $(a, b) \in R \cup S$ then $(a, b) \in R$ or $(a, b) \in S$. If $(a, b) \in R$, since R is symmetric, $(b, a) \in R$ so $(b, a) \in R \cup S$. A similar argument suffices if $(a, b) \in S$, so $R \cup S$ is symmetric. Now let $(a, b), (b, c) \in R \cup S$. If both are in R or both are in S, the transitivity of each implies that $R \cup S$ is transitive, so suppose $(a, b) \in R$ and $(b, c) \in S$. Since $R \circ S = S \circ R$ and $(a, c) \in S \circ R$, $(a, c) \in R \circ S$. Thus there exists a $d \in A$ such that $(a, d) \in S$ and $(d, c) \in R$. But both R and S are symmetric so $(c, d) \in R$ and $(d, a) \in S$. Therefore $(c, a) \in R$. But R is symmetric so we have $(a, c) \in R$ and hence $R \cup S$ is transitive. A similar argument can be used if $(a, b) \in S$ and $(b, c) \in R$. $\qquad\square$

    "Counterexample": Let $A = \{a, b, c, d\}$,

    $$R = I_A \cup \{(a, b), (b, a), (a, c), (c, a)\},$$

    $$S = I_A \cup \{(c, d), (d, c), (a, c), (c, a), (d, a), (a, d)\}.$$

    Then R, S are equivalence relations with $R \circ S = S \circ R$ but $R \cup S$ contains (b, a) and (a, d) but not (b, d) and thus is not transitive.

13. **Believe It or Not:** Conjecture: Let R be a strict total order on a non-empty set A. Then $S = A \times A - R$ is a total order on A.

    "Proof": Since R is irreflexive, $I_A \cap R = \varnothing$ so $I_A \subseteq S$ and S is reflexive. Now suppose that $(a, b), (b, c) \in S$. R is complete, so since

$(a, b), (b, c) \in S$, we have $(b, a), (c, b) \in R$. The transitivity of R implies that $(c, a) \in R$. If $(a, c) \in R$ then (a, a) is also in R, which is impossible; hence, $(a, c) \in S$ and S is transitive. Now suppose that $(a, b), (b, a) \in S$. Then we must have $a = b$; otherwise R would not be complete. Thus S is a partial order. If $a, b \in A, a \neq b$ and $(a, b) \notin S$, then $(a, b) \in R$, and as we noted before, this implies $(b, a) \in S$ and hence S is complete and thus a total order. $\qquad \square$

"Counterexample": Let $A = \{1, 2, 3\}, R = \{(1, 2), (2, 3), (3, 1)\}$. Then

$$S = \{(1, 1), (2, 2), (3, 3), (2, 1), (1, 3), (3, 2)\}$$

is not a total order because it is not transitive ($(1, 3), (3, 2) \in S$ but $(1, 3) \notin S$).

2.5 EQUIVALENCE RELATIONS AND PARTITIONS

Let us take a closer look at the equivalence relation R which was example 11 in section 2.3 (recall that R was the relation on $\mathbb{N}$ given by xRy if and only if $5 \mid (x - y)$). If we define $S_i = \{x : xRi\}$ we see that

$$S_1 = \{x : xR1\} = \{1, 6, 11, 16, \ldots\},$$
$$S_2 = \{x : xR2\} = \{2, 7, 12, 17, \ldots\},$$
$$S_3 = \{x : xR3\} = \{3, 8, 13, 18, \ldots\},$$
$$S_4 = \{x : xR4\} = \{4, 9, 14, 19, \ldots\},$$
$$S_5 = \{x : xR5\} = \{5, 10, 15, 20, \ldots\},$$
$$S_6 = \{x : xR6\} = \{1, 6, 11, 16, \ldots\} = S_1,$$
$$S_7 = S_2, \quad S_8 = S_3, \text{ etc.}$$

or more graphically:

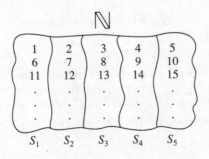

There are several interesting things to note about these sets. While at first glance one might suppose that there are an infinite number of them, there are only five. Also, the union of these five sets is all of $\mathbb{N}$; that is, given any element $y \in \mathbb{N}$, y is an element of one of these five sets. To be more precise, there is exactly *one* of these sets which has y as an element. This also means (see exercise 2 below) that, given any two of these sets, either the sets are equal or they are disjoint. We will see shortly that every equivalence relation generates sets with these properties; but first we need some definitions.

Definition 2.13: Let A be a non-empty set. A *partition* Π of A is a collection of non-empty subsets of A such that every element of A is an element of exactly one of these sets.

We note that if Π is a partition of A, then the elements of Π are subsets of A; we will call them the *blocks* of Π. We see that if B and C are blocks of Π, then either $B = C$ or $B \cap C = \varnothing$. Also the union of all the members of Π is A. Thus we can think of a partition of a set as a cutting-up of the set into disjoint pieces; e.g., we might consider the four classes of an undergraduate college (freshman, sophomore, junior, senior) as a partition of the student body. Here are some examples of partitions:

Examples

1. Let A be non-empty. Then $\Pi_1 = \{\{x\} : x \in A\}$ and $\Pi_2 = \{A\}$ are partitions of A. In a certain sense, Π_1 is the "finest" partition of A while Π_2 is the "coarsest" (see exercise 3 below).
2. Let $A = \{1, 2, 3, 4\}$. Then $\Pi_1 = \{\{1\}, \{2, 3\}, \{4\}\}$ and $\Pi_2 = \{\{1, 4\}, \{2, 3\}\}$ are partitions of A.
3. Referring to the the sets S_i mentioned at the beginning of this section we see that

$$\{S_1, S_2, S_3, S_4, S_5\}$$

is a partition of $\mathbb{N}$.

There is a very close relationship between partitions and equivalence relations; in fact, given an equivalence relation on a set we can generate a partition (as we did for $\mathbb{N}$ above) and given a partition we can generate an equivalence relation. To see how this works we need one more definition:

Definition 2.14: Let R be an equivalence relation on a non-empty set A. Let $x \in A$. The *equivalence class of x modulo R*, denoted by $[x]_R$ (or sometimes x/R), is the set of all elements of A which are R-related to x. In symbols

$$[x]_R = \{y \in A : yRx\}.$$

The set of all such equivalence classes is denoted by $[A]_R$ (or sometimes A/R) and is called A *modulo R*. In symbols

$$[A]_R = \{[x]_R : x \in A\}.$$

Referring once again to the example at the beginning of this section we have

$$[2]_R = S_2 = \{2, 7, 12, 17, \ldots\},$$

$$[4]_R = [9]_R = [14]_R,$$

$$[\mathbb{N}]_R = \{[1]_R, [2]_R, [3]_R, [4]_R, [5]_R\}$$

$$= \{[6]_R, [12]_R, [18]_R, [9]_R, [25]_R\}.$$

For two extreme examples, let A be non-empty and let R be equality (i.e., xRy if and only if $x = y$) and let $S = A \times A$ (also an equivalence relation). Then

$$\forall x \in A, [x]_R = \{x\} \text{ and } [x]_S = A,$$

$$[A]_R = \{\{x\} : x \in A\},$$

$$[A]_S = \{A\}.$$

In the case of R we have each equivalence class containing exactly one element while in the case of S there is just one big equivalence class consisting of all of A.

Perhaps by now you have guessed that $[A]_R$ is a partition of A, with the equivalence classes forming the blocks of the partition. That this is indeed the case we shall shortly prove, but first we will establish a few properties of equivalence classes.

Theorem 2.9: Let R be an equivalence relation on a non-empty set A. Then

a) $\forall x \in A, [x]_R \neq \emptyset$.

b) $\forall x, y \in A, [x]_R \cap [y]_R \neq \emptyset$ if and only if xRy.

c) $\forall x, y \in A, [x]_R = [y]_R$ if and only if xRy.

d) $\forall x, y \in A, [x]_R \neq [y]_R$ if and only if $[x]_R \cap [y]_R = \emptyset$.

Proof: a) Since R is reflexive, xRx for all $x \in A$ so $x \in [x]_R$ and thus $\forall x \in A, [x]_R \neq \varnothing$. b) Suppose that x, y are elements of A and that $[x]_R \cap [y]_R \neq \varnothing$. Let $z \in [x]_R \cap [y]_R$. Then xRz and yRz. Since R is symmetric, zRy and as R is also transitive we have xRy as desired. Now suppose that xRy. Then $y \in [x]_R$. But $y \in [y]_R$ too, so $[x]_R \cap [y]_R \neq \varnothing$. c) If $[x]_R = [y]_R$ then $y \in [x]_R$ so xRy. Now suppose that xRy and let $z \in [x]_R$. Then xRz. By the symmetry of R we have yRx and by R's transitivity we get yRz so $z \in [y]_R$ and hence $[x]_R \subseteq [y]_R$. A similar argument (just interchange the roles of x and y) shows that $[y]_R \subseteq [x]_R$, which completes the proof. d) This follows directly from parts b) and c). $\square$

We have now done most of the work in showing that $[A]_R$ is a partition of A; about all that remains is to write out the proof.

Theorem 2.10: Let A be a non-empty set and R an equivalence relation on A. Then $[A]_R$ is a partition of A.

Proof: We must show that $[A]_R$ is a collection of non-empty subsets of A which has the property that each $x \in A$ is an element of exactly one of the sets of $[A]_R$. As $[A]_R = \{[x]_R : x \in A\}$ and $x \in [x]_R$ for every $x \in A$, each member of $[A]_R$ is non-empty. This also shows that each $x \in A$ is an element of at least one of the sets, namely its own equivalence class. Suppose that there exists a $y \in A$ such that y is an element of two distinct sets in $[A]_R$. But our previous theorem has shown that distinct elements of $[A]_R$ are disjoint which contradicts the fact that y is in both. Therefore, each element of A is an element of exactly one of the equivalence classes. $\square$

Thus we see that an equivalence relation on a set induces a partition of the set. It turns out that this process also works in reverse; i.e., a partition of a set induces an equivalence relation on the set. Before we show this we need a name for such a relation.

Definition 2.15: Let Π be a partition of a set A. We define a relation A/Π (read "A modulo Π") on A by $(x, y) \in A/\Pi$ if and only if there exists a $B \in \Pi$ such that $\{x, y\} \subseteq B$. In words, x is related to y if and only if x and y are both elements of the same block of the partition.

Referring to Π_2 in example 2 given earlier in this section, we have

$$A/\Pi_2 = \{(1, 1), (1, 4), (2, 2), (2, 3), (3, 2), (3, 3), (4, 1), (4, 4)\}.$$

A quick check reveals that this is an equivalence relation, although, of course, we must prove that this is always the case.

Theorem 2.11: Let Π be a partition of a non-empty set A. Then A/Π is an equivalence relation on A.

Proof: Let Π be a partition of A and for notational convenience let's write A/Π as R. We must show that R is reflexive, symmetric and transitive. Let $x \in A$. As x is an element of some block of Π, we have xRx so R is reflexive. If xRy then x and y are both in the same block of Π so clearly y and x are in the same block of Π which implies yRx so R is symmetric. Now suppose that xRy and yRz. Then there exist $B, C \in \Pi$ such that $\{x, y\} \subseteq B$ and $\{y, z\} \subseteq C$. Now $y \in B \cap C$ so $B \cap C \neq \emptyset$ and thus $B = C$. Therefore $\{x, z\} \subseteq B$ and xRz which makes R transitive and hence an equivalence relation. $\qquad\square$

We have now come full circle; a partition Π induces an equivalence relation A/Π and an equivalence relation R induces a partition $[A]_R$. It turns out that the partition induced by an equivalence relation induces the original partition and conversely. In symbols

$$[A]_{A/\Pi} = \Pi \text{ and } A/[A]_R = R.$$

The proof of this interesting fact will be left to the exercises.

Exercises 2.5

1. Let $A = \{1, 2, 3, 4, 5, 6\}$ and $\Pi = \{\{2, 4, 6\}, \{1, 5\}, \{3\}\}$. List the elements of A/Π. Find $[2]_{A/\Pi}$.

2. Let Π be a partition of A and let $B \in \Pi$, $C \in \Pi$. Show that if $B \cap C \neq \emptyset$ then $B = C$.

3. Let Π_1, Π_2 be partitions of A. We say Π_1 is *finer* than Π_2 and write $\Pi_1 \preceq \Pi_2$ if and only if $\forall B \in \Pi_1, \exists C \in \Pi_2 \ni B \subseteq C$.
   a) If $A = \{1, 2, 3, 4\}$ give examples of partitions Π_1, Π_2 such that
      i) $\Pi_1 \preceq \Pi_2$.
      ii) Π_1 is not finer than Π_2 and Π_2 is not finer than Π_1.
   b) Let Π_1, Π_2 be as in example 1 of this section and let Π be any other partition of A. Show that $\Pi_1 \preceq \Pi \preceq \Pi_2$.

4. Let R be an equivalence relation on A. Show that $A/[A]_R = R$.

5. Let Π be a partition of A. Show that $[A]_{A/\Pi} = \Pi$.

6. Let R_1, R_2 be equivalence relations on A. We say that R_1 is *finer* than R_2 and write $R_1 \leq R_2$ if and only if $R_1 \subseteq R_2$.
   a) Let $A = \{1, 2, 3, 4\}$. Give examples of equivalence relations R_1, R_2 such that:
      i) $R_1 \leq R_2$.
      ii) R_1 is not finer than R_2 and R_2 is not finer than R_1.
   b) Let A be a non-empty set and let $\Omega = \{R : R$ is an equivalence relation on $A\}$. Show that $\leq$ is a partial order on Ω. What can be said about whether $\leq$ is complete or not?
   c) If R_1 and R_2 are equivalence relations on a non-empty set A with $R_1 \leq R_2$, is the partition induced by R_1 finer than the partition induced by R_2, vice-versa, or neither?

7. Let Ψ and Π be partitions of a non-empty set A. We define

$$\Psi \star \Pi = \{C \cap D : C \in \Psi, D \in \Pi, C \cap D \neq \varnothing\}.$$

   a) Let $A = \{1, 2, 3, 4, 5\}$, $\Psi = \{\{1, 2, 3\}, \{4, 5\}\}$, $\Pi = \{\{1, 2\}, \{3, 4\}, \{5\}\}$. Find $\Psi \star \Pi$.
   b) Show that if Ψ and Π are partitions of a non-empty set A then $\Psi \star \Pi$ is a partition of A.
   c) Show that $\Psi \star \Pi$ is finer than Ψ and Π.

8. We will generalize the equivalence relation given as example 11 in section 2.3 and discussed at the beginning of this section. Let $m \in \mathbb{N}$. If $x, y \in \mathbb{Z}$ we say $x \equiv y(\text{mod } m)$ if and only if $m \mid (x - y)$. [Note: $x \equiv y(\text{mod } m)$ is read as "x is congruent to y modulo m."] Thus the equivalence relation mentioned earlier is congruence modulo 5. One final bit of notation: we will write the equivalence classes of congruence modulo m as $[x]_m$ and denote the set of all equivalence classes modulo m by $\mathbb{Z}_m$. Thus $\mathbb{Z}_5 = \{[1]_5, [2]_5, [3]_5, [4]_5, [5]_5\}$.
   a) Find $[3]_3, [2]_3, [5]_1$.
   b) Find two solutions to each of the following:
      i) $x \equiv 3 \ (\text{mod } 14)$.
      ii) $x^2 \equiv 2 \ (\text{mod } 7)$.
      iii) $x^2 \equiv 3 \ (\text{mod } 7)$.
   c) Let $m, n \in \mathbb{N}$. Show that if $m \mid n$ then $\mathbb{Z}_n$ is finer than $\mathbb{Z}_m$.
   d) Let $m \in \mathbb{N}$. Show that $\forall x, y, z \in \mathbb{Z}, x \equiv y(\text{mod } m)$ implies $x + z \equiv y + z(\text{mod } m)$ and $xz \equiv yz(\text{mod } m)$.

9. Let R and S be equivalence relations of a non-empty set A. We know that $R \cap S$ is also an equivalence relation on A.
   a) Let $x \in A$. Show that $[x]_{R \cap S} = [x]_R \cap [x]_S$.
   b) Show that $[A]_{R \cap S} = [A]_R \star [A]_S$, where $\star$ is the operation defined in exercise 7.

10. If $p, q \in \mathbb{N}$, we know from exercise 7 that $\mathbb{Z}_p \star \mathbb{Z}_q$ is a partition of $\mathbb{Z}$. Can an $n \in \mathbb{N}$ be found such that $\mathbb{Z}_p \star \mathbb{Z}_q = \mathbb{Z}_n$? If so, prove your result; if not, give a counterexample to show that this partition is not of this form.

11. ***Believe It or Not:*** Conjecture: Let A be a non-empty set and Π, Ψ be partitions of A. If $\Pi \leq \Psi$ and $\Psi \leq \Pi$ then $\Pi = \Psi$.

    "Proof": Let Π, Ψ be as above and let $B \in \Pi$. Since $\Pi \leq \Psi$, there exists a $C \in \Psi$ such that $B \subseteq C$. But since $\Psi \leq \Pi, C \subseteq B$, and hence $B = C$ so $B \in \Psi$. A similar argument shows that $\Psi \subseteq \Pi$ and hence we have $\Pi = \Psi$. $\qquad\square$

    "Counterexample": Let $A = \{1, 2, 3, 4, 5\}$ with

    $$\Pi = \{\{1, 2\}, \{3\}, \{4, 5\}\},$$

    $$\Psi = \{\{1\}, \{2, 3, 4\}, \{5\}\}.$$

    Then $\Pi \leq \Psi$ ($\{3\} \subseteq \{2, 3, 4\}$) and $\Psi \leq \Pi$ ($\{1\} \subseteq \{1, 2\}$) but clearly $\Pi \neq \Psi$.

2.6 FUNCTIONS

One of the most prevalent ideas in mathematics is that of function. No doubt you have been exposed to functions in high school mathematics and functions play a major role in both precalculus and calculus. But although they are familiar objects in your mathematical repertoire, you might be hard-pressed to give a precise definition of them. We will remedy this situation immediately and discover that functions are like grandparents—they are special relations.

Definition 2.16: Let f be a relation from A to B. Then f is a *function* from A to B (denoted by $f : A \to B$, read as "f is a function from A to B") if and only if

a) $Dom(f) = A$.
b) $\forall x \in A, \forall y, z \in B, [(x, y) \in f \wedge (x, z) \in f] \to y = z$.

In words, this says that if f is a relation from A to B such that for every $x \in A$ there exists *exactly one* $y \in B$ such that $(x, y) \in f$ then f is a function. Condition a) guarantees that for each $x \in A$ there will be at *least* one such y and condition b) guarantees that there will be at *most* one, so taken together we get the "exactly one."

If f is a function from A to B then the "functional property" of *each* $x \in A$ being related to *exactly one* $y \in B$ enables us to use the familiar functional notation $y = f(x)$. If f were just any old relation then there might be several (or no) elements of B related to each element of A and the notation "$f(x)$" would not refer to an element of B but would have to refer to a subset of B.

As an example of some relations which are functions and some which are not functions, let

$$A = \{1, 2, 3, 4\},$$

$$B = \{1, 2, 3, 4, 5\},$$

$$f = \{(1, 2), (2, 3), (3, 4), (4, 5)\},$$

$$g = \{(1, 2), (1, 3), (2, 4), (3, 5), (4, 5)\},$$

$$h = \{(1, 1), (2, 2), 3, 3)\}.$$

Then f, g and h are all relations from A to B but only f is a function; g is not a function because both $(1, 2)$ and $(1, 3)$ are elements of g and h is not a function because $Dom(h) = \{1, 2, 3\} \neq A$. f has a particularly simple form for we can describe it with a formula: $\forall x \in A, f(x) = x + 1$. While most of the functions from precalculus and calculus were given in a similar formula manner, it is certainly not necessary that functions be described in this fashion; in fact, most functions cannot be specified in such a simple way.

We will use the following notation and names when working with functions:

If $f : A \to B$ and $(x, y) \in f$ then we write $y = f(x)$. Note that the function's name is f and that $f(x)$ is not the function's name but is an element of B: it is that particular element which is related to a certain element of A, namely x. If $y = f(x)$ then we say that y is the *image of x* and x is a *preimage of y*. Also observe that we are justified in using *the* when speaking of the image but must use *a* when talking about preimages as an element of B may have several elements in A related to it. Since f is a relation we can treat it as a relation and speak of its domain and image, compose it with other functions and talk about its inverse. We also note that although $Dom(f) = A$ we need not have $Im(f) = B$ so it will be convenient to have a name for B; we will call it the *codomain* of f.

Consider the following example: let $A = \{1, 2, 3, 4, 5\}$, $B = \{a, b, c, d\}$ and define $f : A \to B$ by $f(1) = b$, $f(2) = b$, $f(3) = a$, $f(4) = d$, $f(5) = a$ (see figure below):

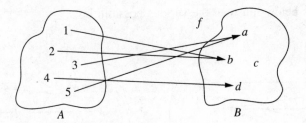

Then the image of 2 is b, a preimage of a is 5 (another preimage of a is 3), c has no preimages.

The following theorem is useful in determining when two functions are equal:

Theorem 2.12: Let $f : A \to B$ and $g : A \to B$. Then $f = g$ if and only if $\forall x \in A, f(x) = g(x)$.

Proof: First, suppose that $f = g$ and let $z \in A$. Then $\exists y \in B \ni (z, y) \in f$. But since $f = g$, $(z, y) \in g$. Thus $y = f(z)$ and $y = g(z)$ so $f(z) = g(z)$.

Now suppose that $\forall x \in A, f(x) = g(x)$. Since functions are relations and relations are sets of ordered pairs, to show $f = g$ we must show that they are equal as sets of ordered pairs. To this end, let $(w, z) \in f$. Then $z = f(w) = g(w)$ so $(w, z) \in g$ and we have $f \subseteq g$. By interchanging the roles of f and g we can show that $g \subseteq f$ from which it follows that $f = g$. $\qquad \square$

There are certain properties which functions may or may not have which come up often enough to deserve names. Some of these are given in

---

Definition 2.17: Let $f : A \to B$. Then:

a) We say f is *one-to-one* (or f is an injection) if and only if $\forall w, z \in A, f(w) = f(z)$ implies $w = z$.

b) We say f is *onto* (or f is a surjection) if and only if $Im(f) = B$.

c) We say f is a *one-to-one correspondence* (or a bijection) if and only if f is both one-to-one and onto.

The figures below illustrate the various possibilities:

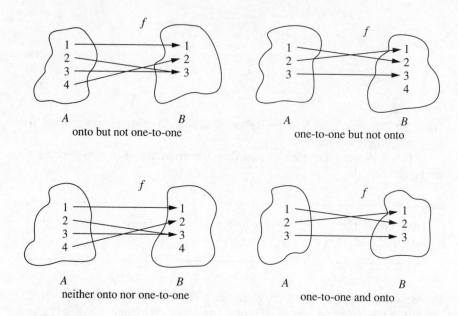

onto but not one-to-one

one-to-one but not onto

neither onto nor one-to-one

one-to-one and onto

Recall that since functions are relations, they have inverses which are relations. Thus, we can talk about the inverse of any function, but there is no reason to expect that a function's inverse will also be a function. It turns out that one-to-one correspondences are particularly important because they are exactly those functions whose inverses are also functions. Here is the theorem and its proof:

Theorem 2.13: Let $f:A \to B$. Then $f^{-1}:B \to A$ if and only if f is a one-to-one correspondence.

Proof: First, suppose that f^{-1} is a function from B to A. We must show that f is one-to-one and onto. Suppose that $f(x) = f(y) = z$. This means that $(x, z) \in f$ and $(y, z) \in f$. Hence $f^{-1}(z) = x$ and $f^{-1}(z) = y$. But f^{-1} is a function so $x = y$ and f is one-to-one. To show that f is onto, let $y \in B$. Then since $Dom(f^{-1}) = B$, there exists an $x \in A$ such that $f^{-1}(y) = x$. Thus $(y, x) \in f^{-1}$ which means that $(x, y) \in f$ so $Im(f) = B$.

Now suppose that f is one-to-one and onto. We must show that f^{-1} is a function from B to A; i.e., we must show that $Dom(f^{-1}) = B$ and that if $(y, x) \in f^{-1}$ and $(y, z) \in f^{-1}$ then $x = z$. First, let $y \in B$. Then since f is onto, there exists an $x \in A$ such that $f(x) = y$ or $(x, y) \in f$. Thus $(y, x) \in f^{-1}$ so $Dom(f^{-1}) = B$. Now suppose that $(y, x) \in f^{-1}$ and $(y, z) \in f^{-1}$. Then $f(x) = y$ and $f(z) = y$. But f is one-to-one so this implies that $x = z$ and hence f^{-1} is a function. $\qquad\square$

It seems to be worth mentioning that it was the one-to-oneness of f which gave f^{-1} the function property and f's ontoness gave us $Dom(f^{-1}) = B$. We should be a little careful about exactly what this theorem says, though. Suppose that $f: A \to B$ and that f is one-to-one but not onto B. Then f^{-1} *is* a function, but *not* a function from B to A; $f^{-1}: Im(f) \to A$.

We actually could have shown a little more about f^{-1} in theorem 2.13, for it is a bijection also. A direct proof of this is possible along the lines above, but to check our understanding of functional notation, consider the following "lazy mathematician's" proof:

Theorem 2.14: If $f: A \to B$ is one-to-one and onto then $f^{-1}: B \to A$ is also a one-to-one correspondence.

Proof: We will use the previous theorem twice: first, it tells us that f^{-1} is a function from B to A. Now interchanging the roles of f and f^{-1} we see that it also tells us that if $(f^{-1})^{-1}$ is a function, then f^{-1} must be one-to-one and onto. But $(f^{-1})^{-1} = f$ so we are done. $\qquad\square$

In exercise 3 below you will get to show that the composition of functions is again a function. Using this fact, if $f: A \to B$ and $g: B \to C$ then $(g \circ f): A \to C$. To see how our functional notation can be used with the composition of functions, we go back to the definition of composition of relations. If $(g \circ f)(x) = z$ then $(x, z) \in (g \circ f)$ which means that there exists a $y \in B$ such that $(x, y) \in f$ and $(y, z) \in g$. Hence, $f(x) = y$ and $g(y) = z$. Therefore $z = g(y) = g(f(x))$ or $(g \circ f)(x) = g(f(x))$, the composition notation we used in calculus. Now you can see why we write composition of relations in the order in which we do: so that when we write composition of functions the order agrees with our usual functional notation. (Digression—some people (especially algebraists) get around this difficulty by changing their functional notation: instead of writing $f(x)$ they write xf—which solves this problem but generates others. Even the invented world of mathematical notation is not free of difficulties!)

Since functions are relations we can compose them and the results we proved about relations will hold. If f, g are functions with suitable domains and images, $(g \circ f)^{-1} = f^{-1} \circ g^{-1}$ although $(g \circ f)^{-1}$, f^{-1} and g^{-1} may not be functions. Theorem 2.14 tells us that if f, g are one-to-one correspondences then f^{-1} and g^{-1} will be functions. To complete the picture we will prove two more theorems about composition and inverses of functions.

Theorem 2.15: Let $f : A \to B$ and $g : B \to C$ be one-to-one correspondences. Then $g \circ f : A \to C$ is a one-to-one correspondence.

Proof: Relying on the results of exercise 3 below we will assume that $g \circ f$ is a function from A to C so we need only show that it is a one-to-one correspondence. First, suppose that there exist $x, y \in A$ such that $(g \circ f)(x) = (g \circ f)(y)$. Thus $g(f(x)) = g(f(y))$. But g is one-to-one so $f(x) = f(y)$. Now the fact that f is one-to-one tells us that $x = y$ so $g \circ f$ must be one-to-one. Next, to show that $g \circ f$ is onto, let $z \in C$. Since g is onto there exists a $y \in B$ such that $g(y) = z$. But f is also onto so there exists an $x \in A$ such that $f(x) = y$. Therefore, $z = g(y) = g(f(x)) = (g \circ f)(x)$ so $g \circ f$ is onto. $\qquad\square$

The above proof is typical of proofs which show that a certain function is one-to-one and onto. To make sure we understand the form of this sort of proof, let's take a closer look. Suppose that $f : A \to B$. A direct proof to show that f is one-to-one would take the form: Let $x, y \in A$, with $f(x) = f(y)$.

$$\cdots$$

"something or other depending on f"

$$\cdots$$

so $x = y$ and f is one-to-one.

A contrapositive proof would be of the form: Let $x, y \in A$, with $x \neq y$.

$$\cdots$$

"something or other depending on f"

$$\cdots$$

so $f(x) \neq f(y)$ and f is one-to-one.

A direct proof to show that f is onto would look like: Let $y \in B$.

$$\cdots$$

"something or other depending on f"

$$\cdots$$

so there exists $x \in A$ such that $f(x) = y$ and f is onto.

To summarize: to show that f is one-to-one we must show that distinct elements in the domain have distinct images and to show that f is onto we must show that every element of B has a preimage.

As an example, let us show that $f: \mathbb{R} \to \mathbb{R}$ given by $f(x) = ax + b$, $a \neq 0$ is a one-to-one correspondence. First, a direct proof that f is one-to-one. Let $x, y \in \mathbb{R}$ with $f(x) = f(y)$. Then $ax + b = ay + b$ which implies that $ax = ay$. Since $a \neq 0$, we have $x = y$ and f is one-to-one. A contrapositive proof of this would be: Let $x, y \in \mathbb{R}$ with $x \neq y$. Then since $a \neq 0$, $ax \neq ay$ so we have $ax + b \neq ay + b$ so $f(x) \neq f(y)$. To show that f is onto, let $z \in \mathbb{R}$. Then $(z - b)/a$ is also an element of $\mathbb{R}$ (since $a \neq 0$) and

$$f(\frac{z - b}{a}) = a(\frac{z - b}{a}) + b = z - b + b = z,$$

so f is onto. We should note that the choice of $(z - b)/a$ did not come out of thin air; it was the result of solving the equation $f(x) = ax + b = z$ for x.

Theorem 2.16: Let $f: A \to B$ and $g: B \to C$ be one-to-one correspondences. Then $(g \circ f)^{-1}: C \to A$ and $\forall x \in C$, $(g \circ f)^{-1}(x) = (f^{-1} \circ g^{-1})(x) = f^{-1}(g^{-1}(x))$.

Proof: We have already done most of the work so we just need to make some observations. First, since f and g are one-to-one correspondences, $g \circ f$ is also a one-to-one correspondence so $(g \circ f)^{-1}$ is a function from C to A. As relations we know that $(g \circ f)^{-1} = f^{-1} \circ g^{-1}$ and since f^{-1} and g^{-1} are functions too, the result follows. $\square$

We note that the identity relation on A, I_A, is a function from A to A which we will call the *identity function*. Using our functional notation, $I_A(x) = x$ for all $x \in A$. The reason for this name (and the name *inverse*) will become clearer in the next theorem.

Theorem 2.17: Let $f: A \to B$. Then

a) $f \circ I_A = f$.
b) $I_B \circ f = f$.
c) If f is a one-to-one correspondence then $f^{-1} \circ f = I_A$ (or $\forall x \in A$, $f^{-1}(f(x)) = x$) and $f \circ f^{-1} = I_B$ (or $\forall x \in B$, $f(f^{-1}(x)) = x$).

Proof: a) and b) follow easily, for suppose that $x \in A$. Then $(f \circ I_A)(x) = f(I_A(x)) = f(x)$ and $(I_B \circ f)(x) = I_B(f(x)) = f(x)$. For c), we first observe that since f is a one-to-one correspondence, f^{-1} is a function so $f(x) = y$ if and only if $f^{-1}(y) = x$. Now, let $x \in A$ and suppose that we let $f(x) = y$. Then $(f^{-1} \circ f)(x) = f^{-1}(f(x)) = f^{-1}(y) = x = I_A(x)$. For the second assertion, let $x \in B$ and suppose that $f^{-1}(x) = y$. Thus $(f \circ f^{-1})(x) = f(f^{-1}(x)) = f(y) = x = I_B(x)$. $\square$

Exercises 2.6

1. Let $A = \{1, 2, 3, 4, 5, 6\}$ and let $f : A \to A$ be given by

$$f(x) = \begin{cases} x + 1, & \text{if } x \neq 6; \\ 1, & \text{if } x = 6. \end{cases}$$

   a) Find $f(3)$, $f(6)$, $f \circ f(3)$, $f(f(2))$.
   b) Find a preimage of 2, 1.
   c) Show that f is a one-to-one correspondence.

2. Show that $f : \mathbb{R} \to \mathbb{R}$ given by $f(x) = x^3$ is one-to-one and onto while $g : \mathbb{R} \to \mathbb{R}$ given by $g(x) = x^2 - 1$ is neither one-to-one nor onto.

3. Suppose that $f : A \to B$ and $g : B \to C$. Show that $g \circ f : A \to C$.

4. a) Let A, B and $f : A \to B$ be:

$$A = \{1, 2, 3, 4\},$$

$$B = \{1, 2, 3\},$$

$$f = \{(1, 3), (2, 1), (3, 1), (4, 2)\}.$$

   Find $f^{-1} \circ f$.
   b) Let A, B be non-empty sets and $f : A \to B$. Show that $f^{-1} \circ f$ is an equivalence relation on A. (Note that f^{-1} may not be a function.) Also show that $[x]_{f^{-1} \circ f} = \{y : f(x) = f(y)\}$.

5. Let $f : A \to B$. Prove that
   a) f is one-to-one if and only if there exists $g : B \to A$ such that $g \circ f = I_A$.
   b) f is onto if and only if there exists $g : B \to A$ such that $f \circ g = I_B$.
   c) f is onto if and only if $f \circ f^{-1} = I_B$.

6. Let $f : A \to B$ and $g : B \to A$. Show that if $g \circ f = I_A$ and $f \circ g = I_B$ then f is a bijection and $g = f^{-1}$.

7. Let R be an equivalence relation on a non-empty set A. We define a relation α from A to $[A]_R$ by

$$\alpha = \{(x, [x]_R) : x \in A\}.$$

a) Show that $\alpha: A \rightarrow [A]_R$.

b) Show that α is onto.

c) Under what conditions will α be one-to-one?

8. Let $f: A \rightarrow A$. Suppose that we know that f is also an equivalence relation. What else can be said about f?

9. Let $f: A \rightarrow B$, $g: A \rightarrow B$. Prove or give counterexamples for the following conjectures:

a) $f \cup g: A \rightarrow B$.

b) $f \cap g: A \rightarrow B$.

c) $f \cup g: A \rightarrow B$ implies $f = g$.

d) $f \cap g: A \rightarrow B$ implies $f = g$.

10. [To refresh your memory about restrictions, see exercise 12, section 2.3.] Let $f: A \rightarrow B$ and $g: C \rightarrow D$, with $A \cap C = \emptyset$.

a) Show that $f \cup g: A \cup C \rightarrow B \cup D$.

b) Show that $f \cup g|_A = f$, $f \cup g|_C = g$.

11. Let $f: \mathbb{R} \rightarrow \mathbb{R}$ be defined by $f(x) = \sin x$.

a) Show that f is not one-to-one.

b) Show that $f|_{[-\pi/2, \pi/2]}$ is one-to-one.

12. Let A be a non-empty set and let

$$\Psi = \{\phi : \phi \text{ is a partition of } A\}.$$

Recall that $\leq$ (is finer than) is a partial order on Ψ. Let

$$\mathfrak{R} = \{R : R \text{ is an equivalence relation on } A\}.$$

We know there is a one-to-one correspondence between elements of Ψ and $\mathfrak{R}$, so let us denote the equivalence relation associated with the partition θ by R_θ. We define a relation $\sqsubseteq$ on $\mathfrak{R}$ by

$$R_\phi \sqsubseteq R_\theta \text{ if and only if } \phi \leq \theta.$$

a) Show that $\sqsubseteq$ is a partial order on $\mathfrak{R}$ (you may assume that $\leq$ is a partial order on Ψ).

b) Prove (or give a counterexample):

$$R_\phi \sqsubseteq R_\theta \text{ if and only if } R_\phi \subseteq R_\theta.$$

13. Suppose that $f: A \rightarrow B$ and $g: B \rightarrow C$, where A, B and C are non-empty sets. Prove or give a counterexample for the following conjectures:

a) $g \circ f$ is a bijection implies f is one-to-one.

b) $g \circ f$ is a bijection implies f is onto.

c) $g \circ f$ is a bijection implies g is onto.

d) $g \circ f$ is a bijection implies g is one-to-one.

14. Let $f: A \to B$, with R a strict total order on A and S a strict total order on B. We say that f is *monotonic* if and only if $\forall x, y \in A$, xRy implies $f(x) S f(y)$.
    a) With the usual order ($<$) on $\mathbb{R}$, give an example of a function which is monotonic.
    b) With the usual order ($<$) on $\mathbb{R}$, give an example of a function which is not monotonic.
    c) If $f: A \to B$ is monotonic, show that f is one-to-one.

15. **Believe It or Not:** Conjecture: Let $f: A \to B$ and let R be a strict total order on A. We define a relation S on B by

$$xSy \leftrightarrow \exists a, b \in A \ni aRb \wedge f(a) = x, f(b) = y.$$

Then S is a strict partial order.

"Proof": Suppose f and R are as above and S is defined as indicated. Let $x \in B$ with xSx. But this means that there exists $a \in A$ such that $f(a) = x$. Thus aRa, which is impossible since R is irreflexive; therefore, S is irreflexive. Now suppose that $x, y, z \in B$ with xSy, ySz. Then there exist $a, b, c \in A$ such that $f(a) = x, f(b) = y, f(c) = z$ and aRb, bRc. Since R is transitive, aRc and hence xSz, and S is transitive also. $\square$

"Counterexample": Let $A = \{1, 2, 3\}, B = \{1, 2, 3, 4\}$ and $f: A \to B$ be given by $f(1) = 1, f(2) = 1, f(3) = 4$. Suppose

$$R = \{(1, 2), (2, 3), (1, 3)\}.$$

Then $S = \{(1, 2), (1, 4)\}$, which is transitive but not irreflexive.

16. **Believe It or Not:** Conjecture: Let $f: A \to B$ and $g: B \to A$. If $g \circ f = I_A$ then $g = f^{-1}$.

"Proof": Let f, g be as above and let $x \in B$. Suppose that $y \in A$ is such that $(x, y) \in g$. Let $z \in B$ be such that $(y, z) \in f$. Since $g \circ f = I_A, (z, y) \in g$. But $(x, y) \in g$ and g is a function so $x = z$. Thus $(x, y) \in f^{-1}$ so $g \subseteq f^{-1}$. Now suppose that $(x, y) \in f^{-1}$. Then $(y, x) \in f$. Since $g \circ f = I_A, (x, y) \in g$ so $f^{-1} \subseteq g$ and we have $g = f^{-1}$. $\square$

"Counterexample": Let $A = \{1, 2, 3\}, B = \{1, 2, 3, 4\}$ with

$$f = \{(1, 2), (2, 1), (3, 3)\}, g = \{(2, 1), (1, 2), (3, 3), (4, 3)\}.$$

Then $g \circ f = I_A$ but $g \neq f^{-1}$ since $(4, 3) \in g$.

17. **Believe It or Not:** Conjecture: Let $f: A \to B$. If $f^{-1} \circ f = I_A$ then f is one-to-one.

"Proof": Suppose f is as above and $x, y \in A$ with $f(x) = f(y) = z$. Then $f^{-1}(z) = x$ and $f^{-1}(z) = y$. But f^{-1} is a function so $x = y$ and we see that f is one-to-one. $\qquad\square$

"Counterexample": Let $A = \{a, b, c\}, B = \{1, 2, 3\}$ with $f(a) = 1$, $f(b) = 2, f(c) = 2$. Then $f^{-1} = \{(1, a), (2, b), (2, c)\}$ but f is not one-to-one.

2.7 MORE FUNCTIONS

We can extend the idea of function in a natural way from individual elements of the domain to subsets of the domain; that is, $f: A \to B$ can be extended to $f: \mathbb{P}(A) \to \mathbb{P}(B)$.

> **Definition 2.18:** Let $f: A \to B$. If $C \subseteq A$ then we define $f(C) = \{f(x) : x \in C\}$. If $D \subseteq B$ then $f^{-1}(D) = \{x : f(x) \in D\}$. $f(C)$ is called the *image* of C and $f^{-1}(D)$ is called the *preimage* of D.

For example, let $f: A \to B$ where $A = \{1, 2, 3, 4\}$, $B = \{1, 3, 5\}$ and f is given by $f(1) = 1$, $f(2) = 1$, $f(3) = 5$, $f(4) = 5$. Then

$$f(\{1, 3\}) = \{1, 5\}, \qquad f(\{1, 2\}) = \{1\},$$

$$f^{-1}(\{1\}) = \{1, 2\}, \qquad f^{-1}(\{4\}) = \varnothing.$$

Note that we have, so far, just defined some sets: with A, B, f, C, D as above, $f(C)$ is a subset of B and $f^{-1}(D)$ is a subset of A. However, we can use these definitions in a natural way to define some functions (with the same name)

$$f: \mathbb{P}(A) \to \mathbb{P}(B),$$

$$f^{-1}: \mathbb{P}(B) \to \mathbb{P}(A).$$

These two induced set functions have many interesting properties (including the fact that they *are* functions!), most of which will be left to the exercises, but we will prove one here to give the flavor of such proofs.

> **Theorem 2.18:** Let $f: A \to B$ and let $C \subseteq D \subseteq B$. Then
> $$f^{-1}(C) \subseteq f^{-1}(D).$$

Proof: Let $x \in f^{-1}(C)$. Then $x \in A$ and $f(x) \in C$. Since $C \subseteq D$, $f(x) \in D$. But this means that $x \in f^{-1}(D)$. $\qquad\square$

All of our familiar arithmetic operations $(+, -, \cdot, \div)$ are functions, as are our logic connectives $(\vee, \wedge, \rightarrow, \leftrightarrow)$ and our set operations $(\cap, \cup, -)$. We have a special name for functions of this kind:

Definition 2.19: Let A be a set. $\bullet$ is called a *binary operation on A* if and only if

$$\bullet: A \times A \rightarrow A.$$

Thus we see that a binary operation on A associates with each pair of elements of A another element of A. Because of this we depart from our usual functional notation and write

$$a \bullet b = c \text{ instead of } \bullet\,((a, b)) = c.$$

For example, with $+: \mathbb{R} \times \mathbb{R} \rightarrow \mathbb{R}$ (addition of real numbers) we write the familiar $2 + 3 = 5$ instead of $+((2, 3)) = 5$.

There are several properties which a binary operation may or may not have:

Definition 2.20: Let $\bullet$ be a binary operation on a set A. Then:

a) We say $\bullet$ is *commutative* if and only if $\forall a, b \in A, a \bullet b = b \bullet a$.

b) We say $\bullet$ is *associative* if and only if $\forall a, b, c \in A, a \bullet (b \bullet c) = (a \bullet b) \bullet c$.

c) We say that $e \in A$ is an *identity for* $\bullet$ if and only if $\forall a \in A, a \bullet e = e \bullet a = a$.

d) If e is an identity for $\bullet$ and $x \in A$ we say that x is *invertible* if and only if $\exists y \in A \ni x \bullet y = y \bullet x = e$. Such a y is called an *inverse* of x.

e) We say $a \in A$ is *idempotent for* $\bullet$ if and only if $a \bullet a = a$.

For some examples of these properties, $+$ on $\mathbb{R}$ is commutative, associative, has 0 as an identity and every element is invertible (the inverse of x is $-x$). The only idempotent element is 0. On the other hand, $-$ on $\mathbb{R}$ is not commutative, not associative, has no identity and the only idempotent element is 0. $\cup$ on $\mathbb{P}(A)$ for some set A is commutative, associative, $\varnothing$ is an identity and every element is idempotent.

We will prove a few theorems involving binary operations and leave some others for the exercises.

Theorem 2.19: Let $\bullet$ be a binary operation on A. Then there exists at most one identity for $\bullet$.

Proof: Suppose that e and e' are identities for $\bullet$. Then

$$e = e \bullet e' = e'.$$

[Note: The first equality holds since e' is an identity and the second is true because e is an identity.] Thus $e = e'$ and $\bullet$ can have at most one identity.

$\square$

Theorem 2.20: If $\bullet$ is an associative binary operation with identity e on A and $x \in A$, then x has at most one inverse.

Proof: Suppose that $x \in A$ has y and y' as inverses. Then

$$x \bullet y = y \bullet x = e \text{ and}$$

$$x \bullet y' = y' \bullet x = e.$$

Thus

$$y = y \bullet e = y \bullet (x \bullet y') = (y \bullet x) \bullet y' = e \bullet y' = y'. \qquad \square$$

By virtue of the above theorems we can speak of "the" identity and "the" inverse of an element (if such exists).

The last idea we wish to introduce in this section may be somewhat subtle, but any effort spent in understanding it will be repaid later (for example, when one encounters the "fundamental theorem of group homomorphisms").

Theorem 2.21: Let $f : A \to B$. We define a relation R on A by xRy if and only if $f(x) = f(y)$. Then R is an equivalence relation on A and hence $[A]_R$ is a partition of A. We define two new functions:

$$\alpha : A \to [A]_R \text{ by } \alpha(x) = [x]_R,$$

$$f^\star : [A]_R \to B \text{ by } f^\star([x]_R) = f(x).$$

Then $f^\star$ is one-to-one and $f = f^\star \circ \alpha$.

Proof: It may be helpful to keep the "picture" below in mind as we work our way through the proof.

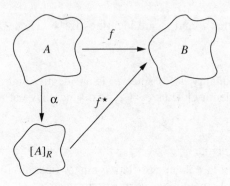

We should observe that there are four things to be proven:

a) R is an equivalence relation.
b) $f^\star$ is a function.
c) $f^\star$ is one-to-one.
d) $f = f^\star \circ \alpha$.

That R is an equivalence relation follows easily from the the fact that $=$ is an equivalence relation on B and the details are left to the exercises. That $f^\star$ is a function is a more subtle point because equivalence classes are involved and $f^\star$ is defined in terms of a representative of the equivalence class. To explain this further, suppose $[x]_R = [y]_R$ with $x \neq y$. For $f^\star$ to be a function and not just a relation we would need $f^\star([x]_R) = f^\star([y]_R)$ (since $[x]_R = [y]_R$) but our definition of $f^\star$ is given in terms of the representatives of the equivalence classes, that is, $f^\star([x]_R) = f(x)$ while $f^\star([y]_R) = f(y)$. Obviously, our only hope to successfully extricate ourselves from this difficulty is to have $f(x) = f(y)$. But if we recall that xRy if and only if $f(x) = f(y)$ we see that $[x]_R = [y]_R$ implies that $f(x) = f(y)$ and $f^\star$ is indeed a function (sometimes we say that such a function defined on a set of equivalence classes is "well-defined"). Thus all that remains to be shown about $f^\star$ is that $f^\star$ is one-to-one. Suppose that $f^\star([x]_R) = f^\star([y]_R)$. Then $f(x) = f(y)$ (from the definition of $f^\star$) so xRy which means that $[x]_R = [y]_R$: therefore, $f^\star$ is one-to-one. To complete the proof we need to show that $f = f^\star \circ \alpha$. Let $x \in A$. Then $(f^\star \circ \alpha)(x) = f^\star(\alpha(x)) = f^\star([x]_R) = f(x)$. $\qquad\square$

To better understand what is going on here, it may help to realize that the equivalence class $[x]_R$ is the set of all elements whose image under f is the same as x's. Thus we can think of the partition $[A]_R$ as "lumping" together all those elements which have the same image under f. As a simple example of this consider: $f: A \rightarrow B$ where $A = \{1, 2, 3, 4\}$, $B = \{1, 3, 5\}$ and f is given by $f(1) = 1$, $f(2) = 1$, $f(3) = 5$, $f(4) = 5$. In this case (you should verify this)

$$R = \{(1, 1), (1, 2), (2, 2), (2, 1), (3, 3), (3, 4), (4, 4), (4, 3)\}$$

so $[1]_R = [2]_R = \{1, 2\}$; $[3]_R = [4]_R = \{3, 4\}$. Hence $[A]_R = \{[1]_R, [3]_R\}$ and $f^\star([1]_R) = 1$, $f^\star([3]_R) = 5$.

Exercises 2.7

1. Let $A = \{1, 2, 3, 4, 5, 6\}$, $B = \{2, 3, 4, 5\}$ and $f: A \rightarrow B$ be given by

   $$f(1) = f(4) = f(6) = 3; f(2) = 5 \text{ and } f(3) = f(5) = 4.$$

   Find:
   a) $f(\{1, 2, 3\})$, $f(A - \{2\})$, $f(A) - \{2\}$.
   b) $f^{-1}(\{3\})$, $f^{-1}(\{4, 5\})$, $f^{-1}(\{2\})$.
   c) $f(\{1, 2\} \cap \{2, 6\})$, $f(\{1, 2\}) \cap f(\{2, 6\})$.

2. Let $f: A \rightarrow B$. Show that:
   a) $C \subseteq D \subseteq A$ implies $f(C) \subseteq f(D)$.
   b) $C \subseteq A$ and $D \subseteq A$ implies $f(C \cup D) = f(C) \cup f(D)$.
   c) $C \subseteq B$ and $D \subseteq B$ implies $f^{-1}(C \cup D) = f^{-1}(C) \cup f^{-1}(D)$.
   d) $C \subseteq B$ and $D \subseteq B$ implies $f^{-1}(C \cap D) = f^{-1}(C) \cap f^{-1}(D)$.
   e) $C \subseteq A$ implies $C \subseteq f^{-1}(f(C))$ and give an example to show that equality need not hold.
   f) $C \subseteq B$ implies $f(f^{-1}(C)) \subseteq C$ and give an example to show that equality need not hold.

3. Let $f: A \rightarrow B$. To distinguish between f and the extension of f to subsets of A let us define f^* to be a relation from $\mathbb{P}(A)$ to $\mathbb{P}(B)$ by

   $$f^* = \{(C, f(C)) : C \in \mathbb{P}(A)\},$$

   and $(f^{-1})^*$ a relation from $\mathbb{P}(B)$ to $\mathbb{P}(A)$ by

   $$(f^{-1})^* = \{(C, f^{-1}(C)) : C \in \mathbb{P}(B)\}.$$

   a) Prove $f^*: \mathbb{P}(A) \rightarrow \mathbb{P}(B)$.
   b) Prove $(f^{-1})^*: \mathbb{P}(B) \rightarrow \mathbb{P}(A)$.
   c) Prove f is one-to-one if and only if f^* is one-to-one.
   d) Prove f is onto if and only if f^* is onto.
   e) When is $f^{-1^*} = (f^*)^{-1}$?

4. Let A be a non-empty set and let $F = \{f : f\colon A \to A\}$. Then $\circ$ (function composition) is a binary operation on F. In answering the following you may want to use some previously proven theorems.
   a) Show that $\circ$ is associative.
   b) Give an example to show that $\circ$ need not be commutative.
   c) Show that I_A is the identity for $\circ$.
   d) Which elements of F have inverses?
   e) Give some examples of idempotent functions.
   f) If f is invertible and $f \circ g = f \circ h$ must $g = h$?
   g) Show that if f and g are invertible then $f \circ g$ is also. What is the inverse of $f \circ g$?

   Perhaps now the names *identity* and *inverse* as used with functions take on more meaning, for I is the identity for $\circ$ and f^{-1} is the inverse of f.

5. Let $\bullet$ be a binary operation on a set A. Show that:
   a) If e is the identity for $\bullet$ then e is idempotent for $\bullet$.
   b) If $\bullet$ is associative and commutative and a, b are both idempotent for $\bullet$ then $a \bullet b$ is also idempotent.
   c) If $\bullet$ is associative and x, y are invertible then $x \bullet y$ is also invertible. Express the inverse for $x \bullet y$ in terms of the inverses of x and y.

6. Let A be a non-empty set. Then $\cup$, $\cap$ and $-$ are binary operations on $\mathbb{P}(A)$. You may wish to quote previously proven theorems and exercises in working the following:
   a) Show $\cup$ and $\cap$ are associative and commutative.
   b) Give examples to show that $-$ is neither associative nor commutative.
   c) Show that every element in $\mathbb{P}(A)$ is idempotent for $\cup$ and $\cap$.
   d) Which elements are idempotent for $-$?
   e) Which elements are invertible for $\cup$, $\cap$, $-$?

7. Let A be a non-empty set. We define a binary operation $\bullet$ on $\mathbb{P}(A)$ by

   $$X \bullet Y = (X - Y) \cup (Y - X).$$

   a) Show that $\bullet$ is commutative.
   b) Show that $\bullet$ is associative.
   c) What is the identity for $\bullet$?
   d) Show that every element in $\mathbb{P}(A)$ is invertible.
   e) If $X \subseteq A$, what is the inverse of X for $\bullet$?

8. Let $F = \{f : f\colon \mathbb{R} \to \mathbb{R}, f(x) = ax + b, a \neq 0\}$. [$F$ is the set of all non-constant linear functions from $\mathbb{R}$ to $\mathbb{R}$.]
   a) Show that $\circ$ (composition of functions) is a binary operation on F.
   b) What is the identity for $\circ$?
   c) Which elements of F are invertible?
   d) If f is invertible, what is the inverse of f?
   e) Which elements of F are idempotent?

9. Suppose that ● is an associative binary operation on A. Let x be a fixed element of A. We define another binary operation $●_x$ on A by

$$a ●_x b = a ● (x ● b).$$

Show that $●_x$ is associative.

10. Show that the relation R in theorem 2.21 is an equivalence relation.

11. Let ● be a binary operation on A. If $B \subseteq A$ we can consider the restriction of ● to B, $●|_B$. This restriction may or may not be a binary operation on B. If $●|_B$ is a binary operation on B we say that B is *closed* with respect to ●.
    a) Give a precise definition of the restriction mentioned above.
    b) Let $+, -$ be the usual algebraic operations on $\mathbb{Z}$. Show that $\mathbb{N}$ is closed with respect to $+$ and not closed with respect to $-$.
    c) If ● is a binary operation on A with $B \subseteq A$, show that B is closed with respect to ● if and only if

$$\{x ● y : x, y \in B\} \subseteq B.$$

12. Let $f: A \rightarrow B$. Show that f can be decomposed into a composition of a surjection, a bijection and an injection; i.e., there exist functions α, β, γ such that $f = \gamma \circ \beta \circ \alpha$ where α is a surjection, β is a bijection and γ is an injection. [Hint: See theorem 2.21.]

13. In ordinary algebra we often use the rule "equals added to equals are equal," or more precisely, if $a, b, c, d \in \mathbb{R}$ with $a = b, c = d$ then $a + c = b + d$. Prove that this is correct.

14. **Believe It or Not:** Conjecture: Let $f: A \rightarrow B$ and $C, D \subseteq A$. Then $f(C \cap D) = f(C) \cap f(D)$.

    "Proof": Let $C, D \subseteq A$ and suppose that $x \in f(C \cap D)$. Then there exists a $y \in C \cap D$ such that $f(y) = x$. Clearly $y \in C$ and $y \in D$, thus $f(y) \in f(C)$ and $f(y) \in f(D)$, so $x \in f(C) \cap f(D)$. Now suppose that $x \in f(C) \cap f(D)$. Then there exists a $y \in C$ such that $f(y) = x$ and there exists a $y \in D$ such that $f(y) = x$. But $y \in C \cap D$ so $x \in f(C \cap D)$. $\square$

    "Counterexample": Let $A = \{1, 2\}, B = \{1, 2, 3\}$ and let $f: A \rightarrow B$ be given by $f(1) = 1, f(2) = 1$. If $C = \{a\}$ and $D = \{2\}$ then $f(C \cap D) = \varnothing$ while $f(C) \cap f(D) = \{1\}$.

15. **Believe It or Not:** Conjecture: Let A, B be sets with $f: A \rightarrow B$. Then $f^{-1} \circ f$ (these are the induced set functions) is an equivalence relation on $\mathbb{P}(A)$.

    "Proof": Let A, B, f be as above and for convenience, let us denote the composition of the induced set functions, $f^{-1} \circ f$, by R. Let $C \in \mathbb{P}(A)$.

Since $f^{-1}(f(C)) = C$, $(C, C) \in R$ and R is reflexive. If $C, D \in \mathbb{P}(A)$, with $(C, D) \in R$ then we have $f^{-1}(f(C)) = D$. Thus,

$$f(D) = f(f^{-1}(f(C))) = f \circ f^{-1}(f(C)) = I_B(f(C)) = f(C).$$

Since $f(C) = f(D)$, $f^{-1}(f(C)) = f^{-1}(f(D))$ so $(D, C) \in R$ and R is symmetric. Now suppose that $(C, D) \in R$ and $(D, E) \in R$. Thus $f^{-1}(f(C)) = D$ and $f^{-1}(f(D)) = E$. Since

$$E = f^{-1}(f(D)) = (f^{-1} \circ f)(f^{-1} \circ f(C))$$

$$= (f^{-1} \circ (f \circ f^{-1}) \circ f)(C) = f^{-1} \circ I_B(f(C)) = f^{-1}(f(C)),$$

and $(C, E) \in R$ so R is transitive. $\qquad\square$

"Counterexample": Let $A = \{1, 2\}$, $B = \{1, 2, 3\}$ and $f : A \to B$ be given by $f(1) = 1$, $f(2) = 1$. Then

$$f^{-1} \circ f = \{(\varnothing, \varnothing), (\{1\}, \{1, 2\}), (\{1, 2\}, \{1, 2\}), (\{2\}, \{1, 2\})\},$$

which is not reflexive or symmetric.

16. Let $Q = \{(m, n) : m, n \in \mathbb{Z} \text{ with } n \neq 0\}$. We define a relation R on Q by

$$(m, n)R(x, y) \text{ if and only if } my = nx.$$

a) Show that R is an equivalence relation.
b) Find three elements in $[(1, 2)]_R$, three elements in $[(1, -1)]_R$.
c) Show that $\forall n \in \mathbb{Z}$, $n \neq 0$, $[(x, y)]_R = [(nx, ny)]_R$.
d) We define a binary operation $\star$ on $[Q]_R$ by

$$[(x, y)]_R \star [(m, n)]_R = [(xm, yn)]_R.$$

Show that this binary operation is "well-defined"; that is, if

$$[(x, y)]_R = [(w, z)]_R \text{ and } [(m, n)]_R = [(p, q)]_R$$

then $[(x, y)]_R \star [(m, n)]_R = [(w, z)]_R \star [(p, q)]_R$.

e) We attempt to define another binary operation on $[Q]_R$ by

$$[(x, y)]_R \oplus [(w, z)]_R = [(x + w, y + z)]_R.$$

Show by example that this "binary operation" is not well-defined, and thus is not actually a binary operation.
f) We try again by defining

$$[(x, y)]_R + [(w, z)]_R = [(xz + yw, yz)]_R.$$

Show that this binary operation is well-defined.

[Note: The alert reader may have made the identification of Q with $\mathbb{Q}$, the set of rational numbers, with (m, n) playing the role of m/n. Actually, what we think of as the number $\frac{1}{2}$ is really an equivalence class and equality of rational numbers is equality of equivalence classes. Now do you see why you wondered about $\frac{1}{2} = \frac{3}{6}$ in the fifth grade?]

17. Let $f: \mathbb{N} \to \mathbb{N}_5$ (see exercise 8, section 2.5 for this notation) be given by $f(x) = [x]_5$. Let R and α be as in theorem 2.21.

   a) Find $[2]_R$, $[9]_R$.

   b) Find $\alpha(4)$, $\alpha(13)$.

   c) We define a binary operation $+$ on $[\mathbb{N}]_R$ by

$$[x]_R + [y]_R = [x + y]_R.$$

   Show that $+$ is indeed a binary operation.

   d) Show that $[5]_R$ is the identity for $+$.

   e) Show that $\forall x, y \in \mathbb{N}, \alpha(x + y) = \alpha(x) + \alpha(y)$.

CHAPTER
3

MATHEMATICAL INDUCTION

3.1 INTRODUCTION

Quite often we wish to prove propositions of the form $\forall n \in \mathbb{N}, P(n)$. For example, we might wish to show that

$$\forall n \in \mathbb{N}, 1 + 2 + 3 + \cdots + n = \tfrac{1}{2}n(n + 1), \tag{1}$$

$$\forall n \in \mathbb{N}, (n - 2)^2 = n^2 - 2n + 4, \tag{2}$$

$$\forall n \in \mathbb{N}, n \text{ is even implies } n^2 \text{ is even.} \tag{3}$$

Propositions (2) and (3) can be easily proven using our fixed-but-arbitrary variable technique (you might try to do this) but proposition (1) does not yield to this method. One reason for this difficulty is that the left-hand side of the equality is not in a closed form and we cannot deal with it algebraically. In fact, for us to even understand what the left-hand side means we have to rely on a certain property of the natural numbers; namely, that given a natural number k there is a "next" natural number, which we call $k + 1$. Thus we might expect that a proof of (1) will involve this "next" property of the natural numbers. This is indeed the case and we will examine in the next section the particular property of $\mathbb{N}$ which enables us to prove propositions of this sort.

3.2 THE PRINCIPLE OF MATHEMATICAL INDUCTION

$\mathbb{N}$, the set of natural numbers, is a familiar mathematical object with which we have been acquainted since our childhood. We know about its various properties—algebraic and otherwise—from experience, but we probably have not thought much about it from an axiomatic point of view. While this is an exciting and rewarding activity, we have other objectives in mind. The interested reader is referred to E. Landau's *Foundations of Analysis*, Chelsea, New York, 1951, for a nice axiomatic treatment, which begins with just five postulates for the natural numbers (the so-called Peano postulates) and in a logical tour-de-force proceeds to build the wonderful structures of the integers, the rational numbers and the real numbers. Here we are concerned with the following axiom, which is the fifth of the five Peano postulates for the natural numbers:

Axiom 3.1: Principle of mathematical induction (PMI): Let S be a subset of $\mathbb{N}$ with the property that

a) $1 \in S$.
b) $\forall k \in \mathbb{N}, k \in S \rightarrow k + 1 \in S$.

Then $S = \mathbb{N}$.

In words, this axiom tells us that if we have a set of natural numbers which contains 1 and every time any natural number is in the set, the next natural number is also in the set, then our set contains all the natural numbers. This property is intuitively appealing, for if 1 is in S then the next number, 2, must be in S. But if 2 is in S then 3 must be in S and so forth, implying that all the natural numbers are elements of S. Of course, it is the *and so forth* which cannot be proven, so this principle (which we will call *the principle of mathematical induction*) must be taken as an axiom; that is, an *assumed* property of the set of natural numbers.

We can use the principle of mathematical induction to prove a proposition of the form $\forall n \in \mathbb{N}, p(n)$ by letting S be the set of natural numbers for which p is true; that is,

$$S = \{n \in \mathbb{N} : p(n) \text{ is true}\}.$$

Thus if we can show $p(1)$ is true $(1 \in S)$ and $p(k) \rightarrow p(k + 1)\,(k \in S \rightarrow k + 1 \in S)$ then $S = \mathbb{N}$ or $\forall n \in \mathbb{N}, p(n)$. Hence, proofs using the principle of mathematical induction usually take the following form:

a) Show that $p(1)$ is true (sometimes called the *basis step*).

b) Show that $p(k) \rightarrow p(k + 1)$ (sometimes called the *induction step*).

As an example, consider this classic theorem, frequently associated with an amusing story (ask your instructor about it) involving the famous mathematician Gauss as a young lad:

$$\forall n \in \mathbb{N}, 1 + 2 + \cdots + n = \tfrac{1}{2}n(n + 1).$$

Here $p(n)$ is "$1 + 2 + \cdots + n = \tfrac{1}{2}n(n + 1)$," so $p(1)$ is "$1 = \tfrac{1}{2}(1)(1 + 1)$," which is clearly true (basis step completed). To complete the induction step we must show that a certain implication $(\forall k, p(k) \rightarrow p(k + 1))$ is true. We will use our usual direct method of proof for such a proposition: choose a fixed-but-arbitrary natural number k, assume the hypothesis ($p(k)$) is true and deduce the truth of the conclusion ($p(k + 1)$). To get started, let $k \in \mathbb{N}$. Suppose that $p(k)$ is true, that is, $1 + 2 + \cdots + k = \tfrac{1}{2}k(k + 1)$. Then

$$\begin{aligned}
1 + 2 + \cdots + k + (k + 1) &= \tfrac{1}{2}k(k + 1) + (k + 1) \\
&= (k + 1)(\tfrac{1}{2}k + 1) \\
&= \tfrac{1}{2}(k + 1)(k + 2)
\end{aligned}$$

so $p(k + 1)$ is true, which completes the induction step and hence the induction proof. Therefore, we have proved by the principle of mathematical induction that

$$\forall n \in \mathbb{N}, 1 + 2 + \cdots + n = \tfrac{1}{2}n(n + 1).$$

We will give several more examples with less commentary; see if you can detect the general form of the proof and follow the steps involved.

Examples

1. If $x \geq 0$ then $\forall n \in \mathbb{N}, (1 + x)^n \geq 1 + x^n$. When $n = 1$ we have $1 + x \geq 1 + x$, an obviously true statement. Suppose that $x \geq 0, k \in \mathbb{N}$ and $(1 + x)^k \geq 1 + x^k$. Then

$$\begin{aligned}
(1 + x)^{k+1} &= (1 + x)^k(1 + x) \\
&\geq (1 + x^k)(1 + x) \\
&= 1 + x^{k+1} + x + x^k \\
&\geq 1 + x^{k+1},
\end{aligned}$$

which completes the induction step. Therefore, $\forall n \in \mathbb{N}, (1 + x)^n \geq 1 + x^n$. (You might try to see where the fact that $x \geq 0$ was used.)

2. $\forall n \in \mathbb{N}, n^2 \leq n$. When $n = 1$ we have $1^2 \leq 1$, which is true. Now suppose that $k \in \mathbb{N}$ and $k^2 \leq k$. Then

$$(k + 1)^2 \leq k + 1 \text{ implies}$$

$$k^2 + 2k + 1 \leq k + 1 \text{ or}$$

$$k^2 + 2k \leq k \text{ which implies}$$

$$k^2 \leq k,$$

our original hypothesis, which we assumed to be true; thus the proof is complete.

A surprising result, isn't it? Isn't induction marvelous! One can prove the most interesting things. Of course, this result is not true, so something must be wrong with the proof. What we have above is an example of an error commonly made by novice inductors; we hope you felt a little queasy while reading it. A closer examination reveals that in the induction step we *assumed* our *conclusion* and then obtained our hypothesis, a proof form which is never valid. If all the implications could be reversed, we could build a valid proof by reversing the order of the steps, but in this case the last step cannot be reversed ($k^2 \leq k$ does not imply $k^2 + 2k \leq k$). The point to remember is: while we may try "working backwards" from the conclusion to the hypothesis when looking for a method proof, to have a valid proof we must be able to reverse all the implications. You may have encountered a similar situation earlier in your mathematical career when you were asked to prove some trig identities.

Now for another example (correct, this time):

3. $\forall n \in \mathbb{N}, D_x x^n = n x^{n-1}$ (here D_x represents differentiation with respect to x). When $n = 1$ we have $D_x x^1 = 1 x^{1-1} = 1$ which is true. Now suppose that $k \in \mathbb{N}$ and $D_x x^k = k x^{k-1}$. Then

$$D_x x^{k+1} = D_x x x^k = 1 x^k + x k x^{k-1} \text{ (using the product rule)}$$

$$= x^k + k x^k$$

$$= (k + 1) x^k$$

which completes the proof. Thus, $\forall n \in \mathbb{N}, D_x x^n = n x^{n-1}$.

4. For every natural number n, $n^3 - n$ is divisible by 3. In symbols we would write $\forall n \in \mathbb{N}, 3 \mid (n^3 - n)$. Recall that $a \mid b$ if and only if $\exists c \in \mathbb{Z} \ni b = ac$. When $n = 1$ we have $3 \mid (1^3 - 1)$ or $3 \mid 0$ which is true since $0 = 3 \cdot 0$. Now suppose that $k \in \mathbb{N}$ and $3 \mid (k^3 - k)$. This means that there exists an integer, say m, such that $k^3 - k = 3m$. Thus

$$(k + 1)^3 - (k + 1) = k^3 + 3k^2 + 3k + 1 - k - 1$$
$$= (k^3 - k) + 3(k^2 + k)$$
$$= 3m + 3(k^2 + k)$$
$$= 3(m + k^2 + k),$$

so $(k + 1)^3 - (k + 1)$ is clearly divisible by 3, which completes the proof.

The principle of mathematical induction can be generalized in the following way: If $S \subseteq \mathbb{Z}$ with the property that

a) $n_0 \in S$,
b) $\forall n \in \mathbb{Z}, n \geq n_0, n \in S \to n + 1 \in S$, then $\{n \in \mathbb{Z} : n \geq n_0\} \subseteq S$. If n_0 is the smallest element in S, then $S = \{n \in \mathbb{Z} : n \geq n_0\}$

We see that the PMI is a special case of this with $n_0 = 1$. As an example of an application of this, consider

5. $\forall n \in \mathbb{N}, n \geq 13, n^2 < (\frac{3}{2})^n$. Here our basis step is $n = 13$. We note that $13^2 = 169 < 194 < 1594323/8192 = (\frac{3}{2})^{13}$, so our basis step is complete. Now suppose that $n > 13$ and $n^2 < (\frac{3}{2})^n$. Then

$$(n + 1)^2 = \left(1 + \frac{1}{n}\right)^2 n^2$$

$$< (1 + \tfrac{1}{13})^2 n^2$$

$$< \tfrac{3}{2} n^2$$

$$< \tfrac{3}{2}(\tfrac{3}{2})^n = (\tfrac{3}{2})^{n+1},$$

which completes the proof.

Now you can have a chance to practice using the principle of mathematical induction yourself.

Exercises 3.2

1. Prove the following (Note: Some of these are not particularly easy):
   a) $\forall n \in \mathbb{N}, 1^2 + 2^2 + \cdots + n^2 = n(n + 1)(2n + 1)/6$.
   b) $\forall n \in \mathbb{N}, 1^3 + 2^3 + \cdots + n^3 = (\frac{1}{2}n(n + 1))^2$.
   c) $\forall n \in \mathbb{N}, 1 + 3 + 5 + \cdots + (2n - 1) = n^2$.
   d) $\forall n \in \mathbb{N}, 1 + 2^{-1} + 2^{-2} + \cdots + 2^{-n} \leq 2$.

e) $\forall n \in \mathbb{N}, n \geq 2, \forall x, y \in \mathbb{R}, x^n - y^n = (x - y)(x^{n-1} + x^{n-2}y + \cdots + xy^{n-2} + y^{n-1})$.

f) $\forall n \in \mathbb{N}, 2 \mid n(n + 1)$.

g) $\forall n \in \mathbb{N}, 7 \mid (3^{2n+1} + 2^{n+2})$ [Hint: $9 = 7 + 2$].

h) $\forall n \in \mathbb{N}, 11 \mid (8 \cdot 10^{2n} + 6 \cdot 10^{2n-1} + 9)$.

i) $\forall n \in \mathbb{N}, D_x^n x^n = n!$.

j) $\forall n \in \mathbb{N}, 2^n > n$.

k) $\forall n \in \mathbb{N}, \forall a, b \in \mathbb{R}, a > b > 0$ implies $a^n > b^n$.

l) $\forall n \in \mathbb{N}, n^n \geq n!$.

m) $\forall n \in \mathbb{N}, 9 \mid (2 \cdot 10^n + 3 \cdot 10^{n-1} + 4)$.

n) $\forall n \in \mathbb{N}, (1 + 1^{-1})(1 + 2^{-1})(1 + 3^{-1}) \cdots (1 + n^{-1}) = n + 1$.

o) $\forall n \in \mathbb{N}, 3 + 11 + 17 + \cdots + (8n - 5) = 4n^2 - n$.

p) $\forall n \in \mathbb{N}, 1 + 1/2^2 + 1/3^2 + \cdots + 1/n^2 \leq 2 - 1/n$.

q) $\forall n \in \mathbb{N}, \forall a, b \in \mathbb{R}, a \geq 0, b \geq 0, a^n + b^n \geq (a + b/2)^n$.

r) $\forall n \in \mathbb{N}, \forall a \in \mathbb{R}, a \neq 1, 1 + a + \cdots + a^n = (1 - a^{n+1})/(1 - a)$.

s) $\forall n \in \mathbb{N}, 1 \cdot 3 \cdot 5 + 3 \cdot 5 \cdot 7 + \cdots + (2n - 1)(2n + 1)(2n + 3) = n(2n^3 + 8n^2 + 7n - 2)$.

t) $\forall n \in \mathbb{N}, 1/(1 \cdot 3) + 1/(2 \cdot 4) + \cdots + 1/[n(n + 2)] = (3n^2 + 5n)/[4(n + 1)(n + 2)]$.

u) $\forall n \in \mathbb{N}, (1 - \frac{1}{2})(1 - \frac{1}{3}) \cdots (1 - 1/n) = 1/n$.

v) $\forall n \in \mathbb{N}, (1 - 1/2^2)(1 - 1/3^2) \cdots (1 - 1/n^2) = \frac{1}{2}(1 + 1/n)$.

2. Show that for all natural numbers $n, n \geq 2$, there exist non-negative integers a, b such that $n = 2a + 3b$.

3. Find n_0 such that $\forall n \in \mathbb{N}, n \geq n_0, n^2 < (\frac{5}{4})^n$ and prove that your result is correct.

4. Suppose that we define a sequence of numbers (a_n) recursively as follows: let $a_1 = 1$ and for $n \geq 2$, let $a_n = a_{n-1} + 2\sqrt{a_{n-1}} + 1$. Show that $\forall n \in \mathbb{N}, a_n$ is an integer.

5. For $n \in \mathbb{N}$ let $a_n = 1 + 2^{-1} + 3^{-1} + \cdots + n^{-1}$. Show that for each $M \in \mathbb{N}$ there exists an $n \in \mathbb{N}$ such that $a_n > M$.

6. *Believe It or Not*: Conjecture: $\forall n \in \mathbb{N}, n \geq 783, 3n^4 + 15n - 7$ is even.

"Proof": When $n = 783, 3n^4 + 15n - 7 = 1,127,634,377,502$, which is even. Now suppose $n \geq 783$ and $3n^4 + 15n - 7$ is even, say $m \in \mathbb{N}$ is such that $3n^4 + 15n - 7 = 2m$. Then

$$3(n + 1)^4 + 15(n + 1) - 7 = 3(n^4 + 4n^3 + 6n^2 + 4n + 1) + 15n + 15 - 7$$
$$= 3n^4 + 15n - 7 + 12n^3 + 18n^2 + 12n + 18$$
$$= 2(m + 6n^3 + 9n^2 + 6n + 9),$$

which is even. $\qquad \square$

"Counterexample": When $n = 1000, 3n^4 + 15n - 7$ is odd since $3n^4 + 15n$ is clearly divisible by 1000 so when the 7 is subtracted, the result will be odd.

7. ***Believe It or Not***: Conjecture: All multiple-choice questions have the same correct answer. [Note: This result would have been helpful for the SAT; however, it is like many existence theorems in mathematics in that, although it shows there is a "universal answer," it doesn't tell us what it is!]

"Proof": First, we make a definition: we will call $n \in \mathbb{N}$ *predictable* if for every set of n multiple-choice questions, each question in the set has the same correct answer. Next, let $S = \{n \in \mathbb{N} : n$ is predictable$\}$. Clearly, $1 \in S$, since each question in a set with only one question has the same correct answer. Now suppose that $n \in \mathbb{N}$ and n is predictable. Let $T = \{q_1, q_2, \ldots, q_{n+1}\}$ be a set of $n + 1$ multiple-choice questions. Now $T - \{q_1\}$ is a set of n multiple-choice questions and hence $\{q_2, q_3, \ldots, q_{n+1}\}$ all have the same correct answer. Also all the questions in $T - \{q_{n+1}\} = \{q_1, q_2, \ldots, q_n\}$ have the same correct answer. But this means that q_1 has the same correct answer as all the rest, so *all* the questions in T have the same correct answer. Since T was an arbitrary set of $n + 1$ questions, $n + 1$ is predictable and $S = \mathbb{N}$. □

"Counterexample": Recall the last multiple-choice test you took.

3.3 EQUIVALENT FORMS OF THE PRINCIPLE OF MATHEMATICAL INDUCTION

In this section we will discuss two other propositions which are equivalent to the principle of mathematical induction. In some situations one of these forms may be easier to use than the others. The first is known as the *well-ordering principle* (WOP):

Let S be a non-empty subset of $\mathbb{N}$. Then S has a least element; that is, there exists a $y \in S$ such that for all $x \in S, y \leq x$.

The second is known as the *principle of complete induction* (PCI) or *course-of-values formulation of PMI*:

If S is a subset of $\mathbb{N}$ such that:

a) $1 \in S$,
b) $\forall n \in \mathbb{N}, \{1, 2, \ldots, n\} \subseteq S \rightarrow n + 1 \in S$,

then $S = \mathbb{N}$.

While the PCI seems to be closely related to the PMI, the connection between these two and the WOP is not so clear. Since we have assumed the PMI as an axiom we could use it to prove the other two as theorems. What we will show, however, is somewhat stronger; that is, they are equivalent:

$$PMI \rightarrow WOP,$$

$$WOP \rightarrow PCI,$$

$$PCI \rightarrow PMI.$$

This will show that all three are logically equivalent and thus we could have chosen any one of them as our axiom and proved the other two as theorems. To start our program of implications we will assume that the principle of mathematical induction holds and prove:

Theorem 3.1: Let S be a non-empty subset of $\mathbb{N}$. Then S has a least element.

Proof: We will use an indirect proof. Suppose S is a non-empty subset of $\mathbb{N}$ which has no least element. Let S^C be the complement of S; that is, $S^C = \mathbb{N} - S$. We define $T = \{x \in \mathbb{N} : \text{for all } y \leq x, y \in S^C\}$. Then since $1 \in S^C$ (for if $1 \in S$ then 1 would be the least element of S, as $\forall x \in \mathbb{N}, 1 \leq x$), $1 \in T$. Now suppose that $k \in T$. Because of the way T is defined, this means that $1, 2, \ldots, k$ must all be elements of S^C. What can we say about $k + 1$? If $k + 1$ were in S then it would be the least element of S, which is impossible since we are assuming that S has no least element. Therefore, $k + 1 \in S^C$ which implies $k + 1 \in T$. Hence, by the principle of mathematical induction, $T = \mathbb{N}$. This means that $S^C = \mathbb{N}$ which implies $S = \varnothing$, a contradiction. Therefore, S must have a least element. $\square$

Next we will assume that the well-ordering principle holds and prove the principle of complete induction:

Theorem 3.2: Let S be a subset of $\mathbb{N}$ such that:

a) $1 \in S$,
b) $\forall n \in \mathbb{N}, \{1, 2, \ldots, n\} \subseteq S \rightarrow n + 1 \in S$.

Then $S = \mathbb{N}$.

Proof: Suppose S is as above and consider S^C. If $S^C = \varnothing$ then we are done so suppose that S^C is non-empty. Then by the well-ordering principle, S^C has a least element, say y. Now $y \neq 1$ because $1 \in S$. What can we say about $1, 2, \ldots, y - 1$? They all must be elements of S for otherwise one of them would be the least element of S^C rather than y. Thus by condition b) we have $y \in S$, a contradiction. Therefore, S^C must be empty which means that $S = \mathbb{N}$. □

Moving right along, we will finish our program of implications by assuming that the principle of complete induction holds and proving the principle of mathematical induction:

Theorem 3.3: Let S be a subset of $\mathbb{N}$ such that:

a) $1 \in S$,
b) $\forall n \in \mathbb{N}, n \in S \rightarrow n + 1 \in S$.

Then $S = \mathbb{N}$.

Proof: Suppose that S has properties a) and b) above. We will use the PCI to prove that $S = \mathbb{N}$. Since $\forall n \in \mathbb{N}, \{1, 2, \ldots, n\} \subseteq S \rightarrow n \in S$ is an obviously true proposition, we have

$$\forall n \in \mathbb{N}, (\{1, 2, \ldots, n\} \subseteq S \rightarrow n \in S) \wedge (n \in S \rightarrow n + 1 \in S)$$

which implies $\forall n \in \mathbb{N}, \{1, 2, \ldots, n\} \subseteq S \rightarrow n + 1 \in S$. Hence, S satisfies the hypotheses of the PCI and consequently $S = \mathbb{N}$. □

To see how these alternative formulations of the principle of mathematical induction can be used to prove propositions, let's prove the familiar:

$$\forall n \in \mathbb{N}, 1 + 2 + \cdots + n = \tfrac{1}{2}n(n + 1)$$

using the well-ordering principle. First, let $p(n)$ be "$1 + 2 + \cdots + n = \tfrac{1}{2}n(n + 1)$." Let $S = \{n \in \mathbb{N} : p(n) \text{ is false}\}$. Now if $S = \varnothing$ we are done, so suppose that $S \neq \varnothing$. Thus, by the well-ordering principle, S has a least element, say x. Since $p(1)$ is obviously true, $1 \notin S$ so $x \neq 1$. Consider $x - 1$. As $x \neq 1$, $x - 1 \in \mathbb{N}$ and $x - 1 \notin S$ which implies that $p(x - 1)$ is true. Thus we have

$$1 + 2 + \cdots + (x - 1) + x = \tfrac{1}{2}(x - 1)x + x$$
$$= x(\tfrac{1}{2}(x - 1) + 1)$$
$$= \tfrac{1}{2}x(x + 1),$$

or $p(x)$ is true, a contradiction, since $x \in S$ means that $p(x)$ is false. Therefore, S must be empty and we are done.

The steps involved in the above proof were very similar to those in the proof where we used the PMI and in fact the PMI seems the more natural choice for this theorem. For a situation where the WOP is a more natural choice we will prove that $\sqrt{2}$ is an irrational number:

Theorem 3.4: $\sqrt{2}$ is irrational.

Proof: We will proceed indirectly. Suppose that $\sqrt{2}$ is rational; that is, suppose that there exist natural numbers r, s such that $\sqrt{2} = r/s$. Then $S = \{k \in \mathbb{N} : k = n\sqrt{2}$ for some $n \in \mathbb{N}\}$ is a non-empty set of natural numbers (in particular, $s\sqrt{2} = r$ so $r \in S$). By the WOP, S has a least element, say x. Let $y \in \mathbb{N}$ be such that $x = y\sqrt{2}$. Now $y(\sqrt{2} - 1) = x - y$ is a natural number less than y (since $0 < \sqrt{2} - 1 < 1$) so $z = y(\sqrt{2} - 1)\sqrt{2}$ is less than x. But $z = 2y - x$ so $z \in \mathbb{N}$ and $z \in S$. Thus we have a contradiction, having found an element of S smaller than x. Hence S must be empty so $\sqrt{2}$ is irrational. ☐

Here is another example of the use of the WOP to prove a familiar result:

Theorem 3.5: The division algorithm: Let $a, b \in \mathbb{N}$. Then there exist integers q, r such that

$$a = bq + r \qquad \text{with } 0 \le r < b.$$

Proof: Let $a, b \in \mathbb{N}$ and let

$$S = \{a - bk : k \in \mathbb{Z}, a - bk \ge 0\}.$$

We note that $S \ne \varnothing$ since $a = a - b \cdot 0 \in S$. By the WOP, S has a least element, say $r = a - bq$. Clearly, r is an integer and $a = bq + r$, so all that remains to be shown is that $0 \le r < b$. By the definition of S, $r \ge 0$. If $r \ge b$, then

$$a - b(q + 1) = r - b \ge 0,$$

so $r - b$ is an element of S. But $r > r - b$, a contradiction; thus, we must have $r < b$. ☐

As you have probably noticed, the key to a proof using the WOP is in the selection of a set whose least element plays a major role in the proof. Once this is done, the proof is usually easy to follow. One needs insight to make such a selection correctly. This insight comes from experience and much practice so don't be discouraged if such proofs seem difficult at first (meaning the first few years).

As an example of a theorem using the PCI consider:

Theorem 3.6: Let $n \in \mathbb{N}$. Then $n = 1$, n is a prime number or n is a product of prime numbers. (Recall that a prime number is a natural number whose only factors are 1 and itself.)

Proof: If we let $p(n)$ be the statement "$n = 1$ or n is a prime or n is a product of primes" then we wish to show $\forall n \in \mathbb{N}, p(n)$. Let $S = \{n \in \mathbb{N} : p(n)$ is true$\}$. Clearly $1 \in S$. Now suppose that $1, 2, \ldots, k$ are all elements of S and consider $k + 1$. If $k + 1$ is a prime we are done, so suppose that $k + 1$ is not a prime. Since $k + 1$ is not a prime it must have factors less than itself (and greater than 1), say r and s; that is, $k + 1 = r \cdot s$. Now r and s are both elements of S and thus are primes themselves or are products of primes. But then we have written $k + 1$ as a product of primes so $k + 1 \in S$ and by the PCI, $S = \mathbb{N}$. $\qquad \square$

The reason that the PCI was more useful here than the PMI was that the factorization of $k + 1$ did not lead us to k but to some other smaller numbers and using the PMI we would not have had as part of our hypotheses that these numbers were elements of S.

Now for a chance at some more practice with mathematical induction!

Exercises 3.3

1. In the proof of theorem 3.1 sets T and S^C were defined. Show that $T \subseteq S^C$.

2. Show that $\mathbb{Z}$ does not have the well-ordering property; that is, give an example of a non-empty subset of $\mathbb{Z}$ which does not have a least element.

3. Use the WOP to show that $\sqrt{3}$ is irrational. Try the same technique used in the proof of theorem 3.4. Show where this technique would break down if it were to be used to show that $\sqrt{4}$ is irrational.

4. Use the WOP to show that $\sqrt{17}$ is irrational.

5. Prove exercises 1 a) and b) in section 3.2 using the WOP.

6. Prove, using any method you like:
   a) $\forall n \in \mathbb{N}, 1^4 + 2^4 + \cdots + n^4 = n(n+1)(2n+1)(3n^2 + 3n - 1)/30$.
   b) $\forall n \in \mathbb{N}, 1^5 + 2^5 + \cdots + n^5 = n^2(n+1)^2(2n^2 + 2n - 1)/12$.
   c) $\forall n \in \mathbb{N}, 1^6 + 2^6 + \cdots + n^6 = n^7/7 + n^6/2 + n^5/2 - n^3/6 + n/42$.
   d) $\forall n \in \mathbb{N}, 1 \cdot 2 + 2 \cdot 3 + \cdots + n(n+1) = n(n+1)(n+2)/3$.
   e) $\forall n \in \mathbb{N}, 2304 \mid (7^{2n} - 48n - 1)$.
   f) $\forall n \in \mathbb{N}, 1/\sqrt{1} + 1/\sqrt{2} + \cdots + 1/\sqrt{n} \leq 2\sqrt{n} - 1$.
   g) $\forall n \in \mathbb{N}, \forall k \in \mathbb{N}, 1^k + 2^k + \cdots + n^k \leq n^{k+1}$.

7. Let α, β be the solutions of the equation $x^2 - x - 1 = 0$, with $\alpha > 0$. For all $n \in \mathbb{N}$, let $F_n = (\alpha^n - \beta^n)/(\alpha - \beta)$.
   a) Find F_1, F_3, F_4. Note: These numbers are known as the *Fibonacci numbers*.
   b) Show that $\forall n \in \mathbb{N}, F_{n+2} = F_{n+1} + F_n$. Note: You will find this recurrence helpful in the remainder of this problem.
   c) Show that $\forall n \in \mathbb{N}, F_n$ is an integer.
   d) Show that $\forall n \in \mathbb{N}, F_n < \left(\frac{13}{8}\right)^n$.
   e) Show that $\forall n \in \mathbb{N}, F_{n+1}^2 - F_n F_{n+2} = (-1)^n$.
   f) Show that $\forall n \in \mathbb{N}, 2 \mid F_{3n}, 2 \nmid F_{3n+1}, 2 \nmid F_{3n+2}$.
   g) Show that $\forall n \in \mathbb{N}$,

$$\sum_{i=1}^{n} F_i = F_{n+2} - 1.$$

   h) Show that $\forall m, n \in \mathbb{N}, F_m F_n + F_{m+1} F_{n+1} = F_{m+n+1}$.
   i) Suppose we define $S_n = F_1^2 + F_2^2 + \cdots + F_n^2$. Find a closed-form expression for S_n and prove that your result is correct.

8. Suppose that we define a sequence of numbers (r_n) recursively as follows: let $r_1 = 1$, $r_2 = \frac{1}{4}$ and for $n \geq 2$, let

$$r_{n+1} = \frac{r_n r_{n-1}}{r_n + r_{n-1} + 2\sqrt{r_n r_{n-1}}}.$$

Show that $\forall n \in \mathbb{N}, r_n = F_{n+1}^{-2}$ (F_n from the previous exercise).

9. The following were found by the students whose names are noted. You might try to find some similar relations yourself. Show that for all $n \in \mathbb{N}$:
   a) $6400 \mid (9^{2n} - 80n - 1)$ (W. Liebe).
   b) $3 \mid (4^n + 2)$ (S. Junker).
   c) $13 \mid (4^{2n+1} + 3^{n+2})$ (D. DeVeny).
   d) $24 \mid (16^n + 9^{3n-2} - 1)$ (D. Kay).

10. Extend the division algorithm (theorem 3.5) to include the case when $a \leq 0$. Also show that the q, r are unique.

11. We define a sequence (a_n) by $a_1 = a_2 = 1$, and for $n \geq 3$, $a_n = 4a_{n-1} + 5a_{n-2}$. Show that for $n \geq 3$, $a_n = \frac{1}{15}(5)^n + \frac{2}{3}(-1)^{n+1}$.

12. **Believe It or Not**: Conjecture: $\forall n \in \mathbb{N}$, n is a prime or $\exists p, q \in \mathbb{Z} \ni$, $n = 2^p 3^q$.

    "Proof": Clearly the assertion is true when $n = 1$ since $1 = 2^0 3^0$. Now suppose it is true when $n = 1, 2, \ldots, k$. If $k + 1$ is a prime, we are done, so suppose that $k + 1$ is not a prime. Then $k + 1 = ab$, where $1 < a < k + 1$, $1 < b < k + 1$. By the induction hypothesis, $a = 2^p 3^q$ and $b = 2^r 3^s$ for some $p, q, r, s \in \mathbb{Z}$. Thus $k + 1 = 2^{p+r} 3^{q+s}$, which completes the proof. □

    "Counterexample": Consider 25. 25 is not a prime and since $2 \nmid 25$ and $3 \nmid 25$, $25 \neq 2^p 3^q$ for any integers p, q.

CHAPTER
4

CONTINUITY
CAREFULLY
CONSIDERED

4.1 INTRODUCTION

One of the important ideas studied in a typical calculus course is the concept of continuity. In most such courses two important consequences of continuity (the intermediate value theorem and the maximum value theorem) are perhaps stated but not proven, usually with a comment to the effect that such a proof is "beyond the scope of this course and is best left for a more advanced course." Well, that "more advanced course" has arrived and we can practice our newly learned skills for dealing with proofs to establish these and other results about continuity. To do this, we will first develop some of the properties of the real number system and then spend a little time considering sequences before we actually get to continuity. While the material which follows is self-contained, it is expected that the reader has completed at least one year of calculus.

4.2 THE REAL NUMBER SYSTEM

Since calculus usually deals with real-valued functions of a real variable, that is, functions with domain and codomain $\mathbb{R}$, any careful study of such functions must be based on a precise description of the properties of $\mathbb{R}$. There are several ways to do this. One may "build" the real number system from simpler systems by starting with the five Peano axioms for the natural numbers and then construct, in turn, the integers, the rational numbers and then the reals. With this approach one must define each new structure in terms of the previous one and then prove that the new structure has the desired properties, based on how it was obtained from the previous one. In this process, one finds that the rational numbers are "really" equivalence classes of ordered pairs of integers and that real numbers are, in one method of construction, sets of rational numbers. We will not take this approach here (leaving it for an even *more* advanced course) but instead give some axioms for the real number system and from these prove that it has the properties we will need. This method has the advantage of being relatively concise, but its disadvantage is that we will sacrifice the knowledge of what the real numbers "really are," and have to be satisfied with just knowing what they act like.

One of the aspects which makes $\mathbb{R}$ so interesting is that it consists of several interdependent structures. It has the familiar algebraic structure, a strict total order and a distance function with the resultant topology (whatever that is). Our starting point will be the algebraic structure:

Axiom 4.1: $\mathbb{R}$ is a set with two binary operations, $+$ (addition) and $\cdot$ (multiplication) and two distinct elements, 0 and 1, such that:

a) $+, \cdot$ are commutative and associative.
b) 0 is the identity for $+$.
c) 1 is the identity for $\cdot$.
d) $\forall x \in \mathbb{R}, \exists -x \in \mathbb{R} \ni x + (-x) = 0$.
e) $\forall x \in \mathbb{R}, x \neq 0, \exists x^{-1} \in \mathbb{R} \ni x \cdot x^{-1} = 1$.
f) $\forall a, b, c \in \mathbb{R}, a \cdot (b + c) = a \cdot b + a \cdot c$.

Using the terminology of chapter 2, we see that every real number has an additive inverse and all non-zero reals have multiplicative inverses. Quite often we will use juxtaposition for multiplication; i.e., $a \cdot b$ will be written as ab and we may write x^{-1} as $1/x$. These axioms give us the familiar algebraic properties of $\mathbb{R}$ and though we assume that they are indeed familiar, it is worthwhile to prove a few just to see how the axioms can be used. For example we can show:

Theorem 4.1: $\forall a, b, c \in \mathbb{R}$,

a) $a \cdot 0 = 0$.
b) $(-a)b = -(ab)$.
c) $a \neq 0$ and $ab = ac$ implies $b = c$.

Proof: a) Let $a \in \mathbb{R}$. Then

$$a + a \cdot 0 = a \cdot 1 + a \cdot 0 = a(1 + 0) = a \cdot 1 = a.$$

By the uniqueness of identities (see theorem 2.19), since $a \cdot 0$ "acts like" 0, it is 0.

b) Let $a, b \in \mathbb{R}$. Then

$$0 = (a + (-a))b = ab + (-a)b.$$

By the uniqueness of inverses (see theorem 2.20), we have $-(ab) = (-a)b$.

The proof of c) is left as an exercise. $\square$

Now that we have introduced the algebraic properties of $\mathbb{R}$, we will add the order structure. We do this with the following:

Axiom 4.2: We assume that $\mathbb{R}$ has a unique subset, $\mathbb{R}_+$, with the properties:

a) $\forall a, b \in \mathbb{R}_+, a + b \in \mathbb{R}_+$ and $ab \in \mathbb{R}_+$.
b) $\forall a \in \mathbb{R}$, exactly one of the following is true:

$$a \in \mathbb{R}_+, a = 0, -a \in \mathbb{R}_+.$$

$\mathbb{R}_+$ is called the set of *positive* real numbers.

It is easy to see (by induction) that since $1 \in \mathbb{R}_+$, $\mathbb{N} \subseteq \mathbb{R}_+$ (details left for exercise 2). We can use the set of positive reals to define the familiar order relation $<$ on $\mathbb{R}$:

Definition 4.1: Let $a, b \in \mathbb{R}$. We say a is *less than* b, denoted by $a < b$ (or $b > a$), if and only if $b - a \in \mathbb{R}_+$. If $a < b$ or $a = b$ we will say a is *less than or equal to* b, denoted by $a \leq b$ (or $b \geq a$).

In fact, $<$ is a strict total order as we now see.

Theorem 4.2: The relation $<$ defined on $\mathbb{R}$ is a strict total order.

Proof: We must show that $<$ is irreflexive, transitive and complete. Suppose that $a \in \mathbb{R}$ with $a < a$. Then $a - a = 0 \in \mathbb{R}_+$ a contradiction of b) of axiom 4.2; thus $<$ is irreflexive. Now suppose that $a, b, c \in \mathbb{R}$ with $a < b$ and $b < c$. Then we have $b - a \in \mathbb{R}_+$ and $c - b \in \mathbb{R}_+$. Since $\mathbb{R}_+$ is closed under addition, $(b - a) + (c - b) = c - a \in \mathbb{R}_+$ so $a < c$ and $<$ is transitive. Now if $a, b \in \mathbb{R}$ with $a \neq b$ then $a - b \neq 0$. Thus, either $a - b \in \mathbb{R}_+$ or $-(a - b) = b - a \in \mathbb{R}_+$. Hence we have $b < a$ or $a < b$ and $<$ is complete and therefore $<$ is a strict total order on $\mathbb{R}$. $\square$

In addition to being a strict total order on $\mathbb{R}$, $<$ interacts "properly" with $\mathbb{R}$'s algebraic structure:

Theorem 4.3: Let $x, y, z \in \mathbb{R}$. Then:

a) $x < y$ implies $x + z < y + z$.
b) $x < y$ and $0 < z$ implies $xz < yz$.

Proof: a) Suppose that $x, y, z \in \mathbb{R}$ with $x < y$. Then $y - x \in \mathbb{R}_+$. Since $y + z - (x + z) = y - x$, we have $x + z < y + z$ as desired.

b) Suppose $x, y, z \in \mathbb{R}$ with $x < y$ and $0 < z$. Then $z \in \mathbb{R}_+$ (see exercise 3 below) and $y - x \in \mathbb{R}_+$. But $\mathbb{R}_+$ is closed under multiplication so we have $(y - x)z = yz - xz \in \mathbb{R}_+$; therefore, $xz < yz$. $\square$

Another aspect of the order structure which will be very important in our work is the idea of "boundedness." We begin with some definitions:

Definition 4.2: Let $A \subseteq \mathbb{R}$. We say s is an *upper bound* for A if $\forall x \in A, x \leq s$. If A has an upper bound we say A is *bounded above*. We say r is a *lower bound* for A if $\forall x \in A, x \geq r$. If A has a lower bound we say A is *bounded below*. If A is bounded both above and below, we say A is *bounded*. If A is not bounded, we say A is *unbounded*.

For example, if $A = (0, 1]$ then A is bounded and $1, 14, 248\pi$ are some of the upper bounds for A while $-312, -\sqrt{2}, 0$ are some of the lower bounds for A. Note that A contains one of its upper bounds but none of its lower bounds. $\mathbb{N}$ is bounded below but not above while $\mathbb{Z}$ is neither bounded above or below.

If a set is bounded above, it should be clear that it has an infinite number of upper bounds, but not all upper bounds are created equal. In particular, we are interested in the following:

Definition 4.3: Let $A \subseteq \mathbb{R}$. If $s \in \mathbb{R}$ is such that

a) s is an upper bound for A, and
b) if t is any upper bound for A, then $s \leq t$,

we say s is the *least upper bound* or *supremum* of A, denoted by sup(A) (see exercise 8 below for the justification of *the*).

Similarly, we say r is the *greatest lower bound* or *infimum* of A, denoted by inf(A), if

a) r is a lower bound for A, and
b) if t is any lower bound for A, then $r \geq t$.

For example, 1 is the least upper bound for $(0, 1]$ and 0 is the greatest lower bound. Note that least upper bounds and greatest lower bounds of a set A may or may not be elements of A.

A useful characterization of a supremum is given in

Theorem 4.4: Let $A \subseteq \mathbb{R}$. Then $s = \sup(A)$ if and only if

a) $\forall \epsilon > 0, \forall x \in A, x < s + \epsilon$ and
b) $\forall \epsilon > 0, \exists x \in A \ni s < x + \epsilon$.

Proof: First, suppose that $s = \sup(A)$ and $\epsilon > 0$. Let $x \in A$. Since s is an upper bound for A we have $x \leq s < s + \epsilon$; thus condition a) holds. Now suppose that for no $y \in A$ is $s < y + \epsilon$. Hence $\forall x \in A$, $x + \epsilon \leq s$. Therefore, $s - \epsilon$ is an upper bound for A, contradicting the assumption that s is the supremum.

We now assume that s is a real number satisfying conditions a) and b). Suppose that $y \in A$ is such that $s < y$. Then $s + (y - s)/2 < y$, contradicting condition a). Hence s is an upper bound for A. Now let t be any upper bound for A. If $t < s$, then for all $x \in A$,

$$x + (s - t) \le t + s - t = s,$$

a contradiction of condition b). Thus s is the least upper bound of A. $\qquad\square$

That a similar characterization exists for greatest lower bounds is left as exercise 9.

Up to this point, we could have substituted $\mathbb{Q}$, the set of rational numbers, for every occurrence of $\mathbb{R}$ in our axioms and theorems. This will come to an end now as we give the so-called completeness property which distinguishes $\mathbb{R}$ from $\mathbb{Q}$:

Axiom 4.3: If A is a non-empty subset of $\mathbb{R}$ which is bounded above, then A has a least upper bound in $\mathbb{R}$.

Although it may not be immediately obvious, this is the property which enables the real numbers to correspond to all the points on the number line, leaving no gaps (as the rational numbers do). To see this, pick your favorite point on the number line, say α (it's *my* favorite). Then $A = \{x \in \mathbb{R} : x < \alpha\}$ is a non-empty set which is bounded above (by α) and by the completeness axiom it has a least upper bound in $\mathbb{R}$, which in fact is α (see exercise 10 for the verification of this). The corresponding situation in $\mathbb{Q}$ does not guarantee a least upper bound in $\mathbb{Q}$, for if we pick an irrational point, say $\sqrt{3}$, then $A = \{x \in \mathbb{Q} : x < \sqrt{3}\}$ is non-empty and bounded above (by 4, for example) but has no least upper bound in $\mathbb{Q}$ (details are left to exercise 11).

One other idea which will be useful to us is that of distance. There are many ways to define the distance between real numbers, but we will use the "natural" one involving absolute value. First, let's recall the definition of absolute value:

Definition 4.4: Let $x \in \mathbb{R}$. The *absolute value of* x, denoted by $|x|$, is given by:

$$|x| = \begin{cases} x & \text{if } x \ge 0, \\ -x & \text{if } x < 0. \end{cases}$$

We see that $|x|$ is actually a function with domain $\mathbb{R}$ and range $\mathbb{R}_+ \cup \{0\}$. Before we give the definition of distance we will use on $\mathbb{R}$, we will give a general definition of distance:

Definition 4.5: Let $d: A \times A \to \mathbb{R}$. We say that d is a *distance* function on A if

a) $\forall x, y \in A, d(x, y) = d(y, x) \geq 0$.
b) $\forall x, y \in A, d(x, y) = 0$ if and only if $x = y$.
c) $\forall x, y, z \in A, d(x, y) + d(y, z) \geq d(x, z)$.

$d(x, y)$ is called the *distance from x to y*.

As an example, we can define a distance function d on any set by $d(x, y) = 1$ if $x \neq y$ and $d(x, y) = 0$ if $x = y$. This doesn't turn out to be a very "interesting" distance function but it does satisfy the three conditions of the definition. The more interesting distance function we will use in $\mathbb{R}$ is defined using the absolute value function: for $x, y \in \mathbb{R}$, we define the distance from x to y by $|x - y|$. It is easy to see that conditions a) and b) are satisfied, but that c) (known as the *triangle inequality*) is true requires a little work, left as exercise 13.

We are now ready to add our last feature to $\mathbb{R}$, open and closed sets. First some notation:

Definition 4.6: Let $x \in \mathbb{R}$ and $\epsilon \in \mathbb{R}_+$. Then

$$N_\epsilon(x) = \{y \in \mathbb{R} : |x - y| < \epsilon\}$$

is called the ϵ *neighborhood of* x. If $A \subseteq \mathbb{R}$ we say A is *open* if $\forall x \in A, \exists \epsilon > 0 \ni N_\epsilon(x) \subseteq A$. A set $B \subseteq \mathbb{R}$ is called *closed* if $B^C = \mathbb{R} - B$ is open.

For example, we see that $\varnothing, \mathbb{R}$ are both open and closed. Also, an "open" interval (a, b) is open in this sense. To see this, suppose $a, b \in \mathbb{R}$ with $a < b$. Let $x \in (a, b)$. Define ϵ to be the minimum $\{|x - a|, |x - b|\}$. Then $\epsilon > 0$ and if $y \in N_{\epsilon/2}(x)$ we have

$$a < x - \frac{\epsilon}{2} < y < x + \frac{\epsilon}{2} < b$$

so $N_{\epsilon/2}(x) \subseteq (a, b)$ and (a, b) is open.

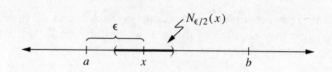

Roughly speaking, for any point x in (a, b) we have enough "room" between it and the nearer endpoint to slip in a neighborhood of x. In the proof above we made our neighborhood extend one-half the way to the nearer endpoint; we could have used $N_\epsilon(x)$ also.

Unions of open sets are open as is the intersection of two open sets. Thus what we have called a closed interval, $[a, b]$, is closed in this sense too, since its complement, $(-\infty, a) \cup (b, \infty)$, is the union of two open intervals. Intervals of the form $(a, b]$ are neither open (there is no neighborhood of b contained in this interval) nor closed (its complement $(-\infty, a] \cup (b, \infty)$ is not open).

This completes our description of the real number system with its algebraic structure, order relation and completeness property, distance function and open sets.

Exercises 4.2

1. Show that $\forall a, b, c \in \mathbb{R}$, $a \neq 0$, $ab = ac$ implies $b = c$.

2. Show that $1 \in \mathbb{R}_+$ and thus that $\mathbb{N} \subseteq \mathbb{R}_+$.

3. Show that $x \in \mathbb{R}_+$ if and only if $0 < x$.

4. Show that $\leq$ is a total order on $\mathbb{R}$.

5. Suppose that $x, y, z \in \mathbb{R}$ with $x < y$ and $z < 0$. Show that $yz < xz$.

6. If x is an upper bound for A and $y \in \mathbb{R}_+$, show that $x + y$ is also an upper bound for A. If x is a lower bound for A and $y \in \mathbb{R}_+$, show that $x - y$ is also a lower bound for A. Give an example of a set which is bounded above but not below.

7. Show that if s is an upper bound for A then $-s$ is a lower bound for $-A = \{-x : x \in A\}$.

8. Show that if $A \subseteq \mathbb{R}$ then A has at most one supremum. Also show that if $\sup(A) = r$, $\inf(A) = s$ and B is a non-empty subset of A then $\sup(B)$ and $\inf(B)$ exist and $s \leq \inf(B) \leq \sup(B) \leq r$.

9. State and prove a characterization for greatest lower bounds similar to theorem 4.4.

10. Let $\alpha \in \mathbb{R}$. Show that α is the supremum of $\{x \in \mathbb{R} : x < \alpha\}$.

11. Let $A = \{x \in \mathbb{Q} : x < \sqrt{3}\}$. Show that A has no least upper bound in $\mathbb{Q}$. [Hint: Suppose $r \in \mathbb{Q}$ is the least upper bound for A. Use the fact that $r < \sqrt{3}$ to find an element of A which is larger than r.]

12. Show that:
a) $\forall x \in \mathbb{R}, |x| \geq 0$.
b) $\forall a \in \mathbb{R}, |-a| = |a|$.
c) $\forall x, y \in \mathbb{R}, x = y$ if and only if $\forall \epsilon > 0, |x - y| < \epsilon$.

13. Show that the triangle inequality holds for $d(x, y) = |x - y|$; that is,
$$\forall x, y, z \in \mathbb{R}, |x - y| + |y - z| \geq |x - z|.$$

14. Show that $N_\epsilon(x) = (x - \epsilon, x + \epsilon)$.

15. Show that if A is a bounded subset of $\mathbb{R}$ with $A \subseteq B$ and B is closed then $\sup(A) \in B$. Give an example to show that if B is not closed the result need not hold.

16. Let $d^*: \mathbb{R} \times \mathbb{R} \rightarrow \mathbb{R}$ be given by $d^*(x, y) = 1$ if $x \neq y$ and 0 otherwise. Show that:
a) d^* is a distance function.
b) Describe the ϵ neighborhoods which this distance function induces.
c) Describe the open sets which arise with this distance function.

17. Show that if $a, b \in \mathbb{R}_+$ then there is an $n \in \mathbb{N}$ such that $na > b$. This is sometimes called the *Archimedean property* of the reals and it shows that no real number is "infinitely small" or "infinitely large." [Hint: Assume this property does not hold, get a bounded set which will have a supremum and then get a contradiction.]

18. *Believe It or Not*: Conjecture: If $a, b \in \mathbb{R}$ with $a < b$ then $1/b < 1/a$.

"Proof": Suppose that $a < b$. Then multiplying both sides by $1/ab$, we obtain

$$\frac{a}{ab} < \frac{b}{ab},$$

or $1/b < 1/a$. $\qquad \square$

"Counterexample": Let $a = -2$ and $b = -1$. Then $a < b$ but $1/a < 1/b$.

19. *Believe It or Not*: Conjecture: If $A \subseteq \mathbb{R}$ with r a lower bound for A and s an upper bound for A then $r \leq s$.

"Proof": Let r be a lower bound for A and s an upper bound. Let $x \in A$. Then we have $r \leq x \leq s$ so $r \leq s$. $\qquad \square$

"Counterexample": Let $A = \emptyset$. Then 1 is an upper bound for A and 2 is a lower bound and clearly $2 \not\leq 1$.

4.3   SEQUENCES

We have already used sequences (in chapter 3, for example) without formally defining them, a deficiency which we will now remedy:

Definition 4.7: A sequence is a function whose domain is $\mathbb{N}$. If a is a sequence and $n \in \mathbb{N}$, we will often denote $a(n)$ by a_n and the sequence by (a_n).

One of the useful features of sequences is the ordering of $\mathbb{N}$, the domain. Thus we think of a_n as the "nth image element." Sometimes we may wish to select certain images from a sequence, say choosing every other one, or using some less regular procedure. We will call this a *subsequence* of the original sequence, a concept more precisely (and also more opaquely) defined in

---

Definition 4.8: Let (a_n) be a sequence and $f : \mathbb{N} \to \mathbb{N}$ a function such that $\forall n \in \mathbb{N}, f(n + 1) > f(n)$. Then $a \circ f$ is a *subsequence of* a. We denote $a(f(k))$ by a_{n_k}.

We can think of a subsequence of (a_n) as an ordered infinite selection of image elements of a. For example, if (a_n) is the sequence given by $a_n = (-2)^n$, $(-2, 4, -8, 16, \ldots)$ then $a_{n_k} = (-2)^{2k}$, $(4, 16, \ldots, 2^{2k}, \ldots)$ is a subsequence of (a_n) while the sequence $4, -2, 16, -8, \ldots$ is not because the original order has been changed.

There are three properties which sequences may or may not have which will be of particular interest to us:

Definition 4.9: Let (a_n) be a sequence of real numbers. If $\{a_n : n \in \mathbb{N}\}$ is bounded we say (a_n) is *bounded*. If $\forall n \in \mathbb{N}, a_{n+1} \geq a_n$ or $\forall n \in \mathbb{N}, a_{n+1} \leq a_n$ we say (a_n) is *monotonic*. If $L \in \mathbb{R}$ we say a_n *converges to* L, denoted by $\lim a_n = L$ or $a_n \to L$, if $\forall \epsilon > 0, \exists M \in \mathbb{N} \ni n \geq M$ implies $|a_n - L| < \epsilon$. If (a_n) converges to something, we say that it is a *convergent* sequence, otherwise we say it is *divergent*.

Thus, for example, (a_n) given by $a_n = 2^n$ is monotonic but not bounded (we say *unbounded*), while (a_n) defined by $a_n = (-1)^n$ is bounded but not monotonic (can you find a monotonic subsequence?). Both of these are divergent sequences. If $\lim a_n = L$ we can think of the elements of

(a_n) as getting and staying arbitrarily close to L as n increases. Thus, if $a_n = 1/n$, then (a_n) is a convergent sequence and $\lim a_n = 0$.

There are several results we wish to prove about convergent sequences, the first of which is that the limit of a convergent sequence is unique:

Theorem 4.5: Let (a_n) be a sequence of real numbers with $\lim a_n = L$ and $\lim a_n = M$. Then $L = M$.

Proof: Suppose that we have a sequence (a_n) with $\lim a_n = L$ and $\lim a_n = M$ with $L \neq M$. Set $\epsilon = |L - M|$. Now there exists $N_1 \in \mathbb{N} \ni n \geq N_1$ implies $|a_n - L| < \epsilon/2$ and there exists $N_2 \in \mathbb{N} \ni n \geq N_2$ implies $|a_n - M| < \epsilon/2$. Thus

$$|L - M| \leq |L - a_{N_1+N_2}| + |a_{N_1+N_2} - M| < \frac{\epsilon}{2} + \frac{\epsilon}{2} = \epsilon,$$

a contradiction. Hence, $L = M$. $\qquad\square$

Note how the triangle inequality was used to make precise the idea that an element of a sequence cannot be arbitrarily close to two distinct numbers. Here is another result about limits of sequences:

Theorem 4.6: Let (a_n) be a sequence of real numbers. If $\{a_n : n \in \mathbb{N}\} \subseteq F$, where F is closed, and $\lim a_n = L$ then $L \in F$.

Proof: Let (a_n) be a sequence of real numbers with $\lim a_n = L$. Suppose that F is a closed subset of $\mathbb{R}$ and $\{a_n : n \in \mathbb{N}\} \subseteq F$ with $L \notin F$. Since L is an element of $\mathbb{R} - F = F^C$ (which is open), there exists $\epsilon_0 > 0$ such that $N_{\epsilon_0}(L) \subseteq F^C$. But $\lim a_n = L$ means that there exists $M \in \mathbb{N}$ such that $n \geq M$ implies $|a_n - L| < \epsilon_0$. Thus $a_M \in F$ and $a_M \in N_{\epsilon_0}(L) \subseteq F^C$, a contradiction. $\qquad\square$

Another intuitively appealing property of sequences is

Theorem 4.7: Let (a_n) be a bounded, monotonic sequence of real numbers. Then (a_n) converges.

Proof: To show that a sequence converges, we must find a candidate for the limit and then show that it is indeed the limit. Let us suppose that (a_n) is monotonically increasing (the decreasing case is left for the exercises).

Since $A = \{a_n : n \in \mathbb{N}\}$ is bounded, let $L = \sup(A)$. We will show that $\lim a_n = L$. Let $\epsilon > 0$. Since L is the supremum of A, there exists an $M \in \mathbb{N}$ such that $L - \epsilon < a_M$. But (a_n) is monotonically increasing, so if $n > M$ then $L - \epsilon < a_n \leq L$. Therefore $\lim a_n = L$. $\square$

And now for our most important (and perhaps most surprising) result about sequences:

Theorem 4.8 (Bolzano-Weierstrass): Let (a_n) be a bounded sequence of real numbers. Then (a_n) has a convergent subsequence.

Proof: Suppose that (a_n) is a bounded sequence. We will construct a subsequence and show that it converges. First, we define a new sequence (b_n) by defining $b_m = \sup(\{a_n : n \geq m\})$, for each $m \in \mathbb{N}$ (since (a_n) is bounded, each of these sets has a supremum). Clearly, (b_n) is a monotonically decreasing, bounded sequence (the verification of these "obvious" facts is left to exercise 7) and hence has a limit, say L. We now construct a subsequence (a_{n_k}) of (a_n) which converges to L. To start, we can find a k such that $b_k \in N_{1/2}(L)$ and hence an $n_1 \geq k$ such that $a_{n_1} \in N_{1/2}(b_k)$. Then, by the triangle inequality, $a_{n_1} \in N_1(L)$. Suppose that we have chosen $n_1 < n_2 < \cdots < n_k$ such that $a_{n_i} \in N_{1/i}(L), 1 \leq i \leq k$. Since (b_n) is decreasing, we can find $r > n_k$ such that $b_r \in N_{1/2(k+1)}(L)$. Then there is an $n_{k+1} \geq r > n_k$ such that $a_{n_{k+1}} \in N_{1/2(k+1)}(b_r)$ and hence $a_{k+1} \in N_{1/(k+1)}(L)$. By the principle of mathematical induction, for all $k \in \mathbb{N}$ we have

$$n_{k+1} > n_k \text{ and } a_{n_k} \in N_{1/k}(L),$$

so (a_{n_k}) is the desired subsequence of (a_n) which converges to L. $\square$

To give a little insight into this interesting result, consider a bounded sequence, say (a_n). If this sequence takes on only a finite number of values (the set $\{a_n : n \in \mathbb{N}\}$ is finite) then we can select a constant subsequence as our convergent subsequence (since at least one value must occur infinitely often). If (a_n) takes on an infinite number of values, then because it is bounded, these values have to "pile up" somewhere; or more precisely, there must be at least one number L such that every neighborhood of L must contain an infinite number of elements of (a_n). It is from these infinite number of elements that we select our subsequence which converges to L. As an illustration of what can happen, suppose that (a_n) is a sequence which consists of all the rational numbers in the interval $[0, 1]$, in some order. Then it is possible to construct a subsequence of (a_n) which converges to *any* real number in $[0, 1]$. Aren't the real numbers fascinating!

Exercises 4.3

1. Let S be a non-empty set of sequences. Show that "is a subsequence of" is a reflexive and transitive relation on S.

2. Show that $\lim a_n = L$ if and only if $\forall \epsilon > 0, \exists M \in \mathbb{N} \ni \{a_n : n \geq M\} \subseteq N_\epsilon(L)$.

3. Write out in detail the negation of $\lim a_n = L$ and prove that $a_n = (-1)^n$ does not converge to any real number L.

4. If we use the "0–1" distance function discussed in exercise 16 of the previous section, describe the convergent sequences.

5. Give an example of a sequence (a_n) and a set A such that $\{a_n : n \in \mathbb{N}\} \subseteq A$ but $\lim a_n \notin A$.

6. Complete the proof of theorem 4.7 by dealing with the case where (a_n) is monotonically decreasing.

7. Show that the sequence (b_n) defined in the proof of theorem 4.8 is in fact monotonically decreasing and bounded.

8. Suppose that (a_n) and (b_n) are sequences of real numbers with $\lim a_n = L$ and $\lim (a_n - b_n) = 0$. Show that $\lim b_n = L$.

9. Suppose that (a_n) is a sequence of real numbers which converges to L. Show that every subsequence of (a_n) converges to L.

10. Give an example of a sequence which has a subsequence which converges to 1, another subsequence which converges to 2 and yet another which is unbounded.

11. If (a_n) is a sequence of real numbers we say it is a *Cauchy sequence* if and only if for all $\epsilon > 0$ there exists $M \in \mathbb{N}$ such that if $m, n \geq M$ then $|a_m - a_n| < \epsilon$. Note that this is an important concept and in the context of $\mathbb{R}$, Cauchy sequences are precisely the convergent sequences.
   a) Give an example of a sequence which is a Cauchy sequence.
   b) Give an example of a sequence which is not a Cauchy sequence.
   c) Show that if (a_n) converges then it is a Cauchy sequence.
   d) Show that if (a_n) is a Cauchy sequence then it is bounded.
   e) Show that if (a_n) is a Cauchy sequence then it converges.

12. *Believe It or Not*: Conjecture: Let (a_n) be a sequence of real numbers. Let $A = \{x \in \mathbb{R} : \text{a subsequence of } (a_n) \text{ converges to } x\}$. Then A is closed.

   "Proof": Let $(a_n), A$ be as above. Let $x \in \mathbb{R} - A$. Then no subsequence of (a_n) converges to x so there exist $M \in \mathbb{N}$ and $\delta_1 > 0$ such that if $n > M$ then $a_n \notin N_{\delta_1}(x)$. Now let $B = \{|a_n - x| : n \leq M\}$ and $\delta_2 = \inf(B)$. Since B is a finite set of positive numbers, $\delta_2 > 0$. If $\delta =$

minimum $\{\delta_1, \delta_2\}$, then $N_\delta(x) \cap A = \emptyset$ so $\mathbb{R} - A$ is open and hence A is closed. $\qquad\square$

"Counterexample": Let (a_n) be defined by $a_n = 1 + (-1)^n - 1/n$, $\forall n \in \mathbb{N}$. Then (a_n) has subsequences which converge to $1, -1$ and 0 so $A = \{-1, 0, 1\}$ which is not closed.

13. **Believe It or Not**: First we make a definition: let $A \subseteq \mathbb{R}$. We say $x \in \mathbb{R}$ is a *limit point of* A if $\forall \delta > 0$, $N_\delta(x)$ contains an element of A other than x $((N_\delta(x) \cap A) - \{x\} \neq \emptyset)$. Conjecture: Let $A \subseteq \mathbb{R}$. Then A is closed if and only if A contains all its limit points.

"Proof": Assume that A is a closed subset of $\mathbb{R}$. Let x be a limit point of A such that $x \notin A$. Then there exists $\delta > 0$ such that $N_\delta(x) \cap A \neq \emptyset$. But this contradicts the assumption that x is a limit point of A so A must contain all its limit points. Now suppose that A contains all its limit points and that $x \notin A$. Then there exists $\delta > 0$ such that $(N_\delta(x) \cap A) - \{x\} = \emptyset$ so $N_\delta \cap A = \emptyset$ and hence A is closed. $\qquad\square$

"Counterexample": Let $A = \{1/n : n \in \mathbb{N}\} \cup \{0\}$. Then the only limit point of A is 0 so A contains all its limit points. But A is not closed because for no $\delta > 0$ is $N_\delta(0) \subseteq A^C$.

4.4 CONTINUOUS FUNCTIONS

Continuous functions form an important class of functions; in fact, in the early days of calculus, all functions were considered to be continuous (or, more correctly, all functions were considered to have the properties of continuous functions, for the precise definition of continuity did not come until later). We start out with a definition of continuity similar to that which you encountered in your calculus class:

Definition 4.10: Let $D \subseteq \mathbb{R}$ and $f : D \to \mathbb{R}$. We say f is *continuous at* $x_0 \in D$ if and only if

$$\forall \epsilon > 0, \exists \delta > 0 \ni \forall x \in D, |x - x_0| < \delta \to |f(x) - f(x_0)| < \epsilon.$$

If $\forall x \in D$, f is continuous at x we say f is *continuous on* D.

Some examples of functions which are continuous throughout their domains are polynomials, rational functions and exponential functions. The traditional example of a function which is not continuous everywhere is the step function (or greatest integer function) f defined by $f(x) = [x]$. This function is not continuous (we say *discontinuous*) at the integers, where its graph makes a "jump."

We could just as well have used neighborhoods in the definition of continuity (the equivalence of these two definitions is left as exercise 2):

Definition 4.10A: Let $D \subseteq \mathbb{R}$ and $f: D \to \mathbb{R}$. We say f is *continuous at* $x_0 \in D$ if and only if

$$\forall \epsilon > 0, \exists \delta > 0 \ni \forall x \in \mathbb{R}, x \in N_\delta(x_0) \cap D \to f(x) \in N_\epsilon(f(x_0)).$$

Another characterization of continuity (which is not so obviously equivalent) involves sequences:

Theorem 4.9: Let $D \subseteq \mathbb{R}$, $f: D \to \mathbb{R}$ and $x_0 \in D$. Then f is continuous at x_0 if and only if for all sequences (a_n) in D which converge to x_0, $(f(a_n))$ converges to $f(x_0)$.

Proof: First, suppose that f is continuous at $x_0 \in D$ and (a_n) is a sequence in D which converges to x_0. Let $\epsilon > 0$. Since f is continuous at x_0, there exists a $\delta > 0$ such that if $x \in D$ and $|x - x_0| < \delta$ then $|f(x) - f(x_0)| < \epsilon$. But $\lim a_n = x_0$ so there exists $M \in \mathbb{N}$ such that $n \geq M$ implies $|a_n - x_0| < \delta$. Thus, if $n \geq M$ we have $a_n \in D \cap N_\delta(x_0)$ so $|f(a_n) - f(x_0)| < \epsilon$ and hence $\lim f(a_n) = f(x_0)$.

Now suppose that for every sequence (a_n) in D which converges to x_0, $(f(a_n))$ converges to $f(x_0)$. We will proceed indirectly by assuming that f is not continuous at x_0. If this is the case, then for some $\epsilon_0 > 0$ and every $\delta > 0$ we can find $x \in D \cap N_\delta(x_0)$ such that $|f(x) - f(x_0)| \geq \epsilon_0$. We will use this assumption to construct a sequence as follows: for each $n \in \mathbb{N}$, let $a_n \in D \cap N_{1/n}(x_0)$ be such that $|f(a_n) - f(x_0)| \geq \epsilon_0$ (the details which verify this are left to exercise 3). Clearly, $\lim a_n = x_0$ but $\lim f(a_n) \neq f(x_0)$ since for all $n \in \mathbb{N}$ we have $|f(a_n) - f(x_0)| \geq \epsilon_0$. This is a contradiction of our hypothesis about the convergence of the sequence of functional values. $\square$

This characterization of continuity using sequences will turn out to be very useful in the near future. It also corresponds (somewhat) to the intuitive approach quite often taken in beginning calculus courses by considering (for example) $f(1.1)$, $f(1.01)$, $f(1.001)$, ..., etc., when examining the continuity of f at 1.

If f is continuous on a set D, we know that for a given $\epsilon > 0$ and for each $x \in D$ there is a $\delta_x > 0$ such that if $z \in D \cap N_{\delta_x}(x)$ then $f(z) \in N_\epsilon(f(x))$. For a given ϵ and some functions and domains no single δ will work for every element of the domain. For example, consider f and

g defined on $(0, 1)$ by $f(x) = x$ and $g(x) = 1/x$. Both are continuous on $(0, 1)$. For $\epsilon = \frac{1}{2}$ we can use $\delta = \frac{1}{2}$ and have $|f(x) - f(z)| < \frac{1}{2}$ whenever $x, z \in (0, 1)$ and $|x - z| < \frac{1}{2}$. But for any $\delta > 0$ we can always find $x, z \in (0, 1)$ with $|x - z| < \delta$ and $|g(x) - g(z)| \geq \frac{1}{2}$ (the details of this are left as exercise 5). Thus some continuous functions and domains are "nicer" than others. The name we use for this is given in

---

Definition 4.11: Let $D \subseteq \mathbb{R}$ and $f : D \to \mathbb{R}$. We say that f is *uniformly continuous on D* if and only if

$$\forall \epsilon > 0, \exists \delta > 0 \ni \forall x, z \in D, |x - z| < \delta \text{ implies } |f(x) - f(z)| < \epsilon.$$

We should note that uniform continuity is a property of both a function *and* its domain. We saw above that $f(x) = 1/x$ is not uniformly continuous on $(0, 1)$ but it is uniformly continuous on, for example, $(1, 2)$. Also, it should be clear that uniform continuity is stronger than continuity; that is, if f is uniformly continuous on D then f is continuous on D (details left as exercise 7). However, if the domain is of the right form, continuity does imply uniform continuity:

Theorem 4.10: Let $f : [a, b] \to \mathbb{R}$ be continuous on $[a, b]$. Then f is uniformly continuous on $[a, b]$.

Proof: Suppose that f is continuous but not uniformly continuous on $[a, b]$. Then there exists some $\epsilon_0 > 0$ such that for every $\delta > 0$ there exist $x, z \in [a, b]$ such that $|x - z| < \delta$ and $|f(x) - f(z)| \geq \epsilon_0$. We will use this to construct a pair of sequences $(x_n), (z_n)$ in $[a, b]$: for each $n \in \mathbb{N}$ let $x_n, z_n \in [a, b]$ with $|x_n - z_n| < 1/n$ and $|f(x_n) - f(z_n)| \geq \epsilon_0$. Now (x_n) is contained in $[a, b]$ so it is bounded and hence has a convergent subsequence, say (x_{n_k}), which converges to something, say x_0. But $[a, b]$ is closed so $x_0 \in [a, b]$. Also (details left as exercise 8), (z_{n_k}) converges to x_0. But f is continuous at x_0 so there exists $\delta > 0$ such that if $y \in N_\delta(x_0) \cap [a, b]$ then $|f(x_0) - f(y)| < \epsilon_0/2$. But for k large enough we have $x_{n_k}, z_{n_k} \in N_\delta(x_0)$ which means that

$$\epsilon_0 \leq |f(x_{n_k}) - f(z_{n_k})| \leq |f(x_{n_k}) - f(x_0)| + |f(x_0) - f(z_{n_k})|$$

$$< \frac{\epsilon_0}{2} + \frac{\epsilon_0}{2} = \epsilon_0,$$

a contradiction. Thus f must be uniformly continuous on $[a, b]$. ☐

Another property of continuous functions on a closed interval is the so-called maximum value property:

Theorem 4.11: Let $f\colon [a, b] \to \mathbb{R}$ be continuous on $[a, b]$. Then f is bounded on $[a, b]$ (i.e., $\{f(x) : x \in [a, b]\}$ is bounded) and there exist $c, d \in [a, b]$ such that for all $x \in [a, b], f(c) \leq f(x) \leq f(d)$; that is, f attains its maximum and minimum values.

Proof: Suppose that f is continuous on $[a, b]$. First, we will show that f is bounded above. To do this, we will proceed indirectly, assuming that f is not bounded above. If this is the case, we can construct a sequence (a_n) in $[a, b]$ such that for all $n \in \mathbb{N}, f(a_n) \geq n$. Since (a_n) is bounded, it has a convergent subsequence, say (a_{n_k}) which converges to an element of $[a, b]$ (why?), say x_0. Now f is continuous at x_0 so $\lim f(a_{n_k}) = f(\lim a_{n_k}) = f(x_0)$. But $f(a_{n_k}) \geq n_k$, a contradiction (for the reason, see exercise 10). Hence f must be bounded above. Since f is bounded above, $M = \sup\{f(x) : x \in [a, b]\}$ exists. We construct another sequence, (b_n), in $[a, b]$, such that for all $n \in \mathbb{N}, |f(b_n) - M| < 1/n$ (why can this be done?). This is a bounded sequence and thus has a convergent subsequence, call it (b_{n_k}), which converges to an element of $[a, b]$, call it d. Now $\lim f(b_{n_k}) = M$ and since f is continuous at d we have

$$f(d) = f(\lim b_{n_k}) = \lim f(b_{n_k}) = M,$$

so that for all $x \in [a, b], f(d) \geq f(x)$. A similar argument (which you can provide in exercise 11) shows that f is bounded below and attains its minimum value. $\square$

The last property of continuous functions on closed intervals which we prove is the intermediate value property:

Theorem 4.12: Let $f\colon [a, b] \to \mathbb{R}$ be continuous on $[a, b]$. If c is any number between $f(a)$ and $f(b)$, then there exists $d \in [a, b]$ such that $f(d) = c$.

Proof: Suppose that f is continuous on $[a, b]$ and that $f(a) < f(b)$ (the case $f(a) \geq f(b)$ is left as exercise 13). Let c be such that $f(a) < c < f(b)$. We define a set S by $S = \{x \in [a, b] : f(x) < c\}$. Then S is non-empty ($a \in S$) and bounded above (by b) and hence has a supremum, call it M (see figure on page 144).

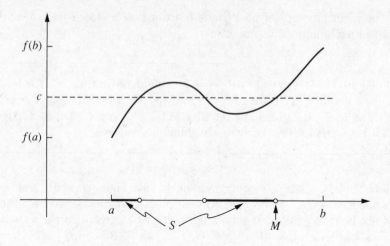

We note that $M \in [a, b]$, so f is continuous at M. We construct two sequences $(a_n), (b_n)$ in $[a, b]$ as follows: for each $n \in \mathbb{N}$, let $a_n \in S$ with $M - 1/n < a_n \leq M$ and let $b_n = \text{minimum } \{b, M + 1/n\}$. Then $\lim a_n = \lim b_n = M$ and since f is continuous at M, $f(M) = \lim f(a_n) = \lim f(b_n)$. But, for all $n \in \mathbb{N}$, $f(a_n) < c$ (since $a_n \in S$) and $f(b_n) \geq c$ (since $b_n \notin S$). Thus we have $f(M) \leq c$ and $f(M) \geq c$ so $f(M) = c$ as desired.   □

This completes our program of proving some of the more useful properties of continuous functions which are usually presented without proof in a beginning calculus course. In working through these proofs and the associated exercises you gained some experience in that part of mathematics which is called *analysis*. As you might expect, we have barely begun to uncover the properties of real-valued functions, to say nothing of the interesting generalizations of these ideas which are possible.

Exercises 4.4

1. Write out a negation of "f is continuous at x_0."

2. Show that definitions 4.10 and 4.10A are equivalent.

3. Verify that the sequence constructed in the proof of theorem 4.9 can actually be constructed.

4. Let $f : \mathbb{R} \to \mathbb{R}$. Show that f is continuous at $x_0 \in \mathbb{R}$ if and only if for all open sets U containing $f(x_0)$, $f^{-1}(U)$ is open.

5. For each $n \in \mathbb{N}$ find $x, z \in (0, 1)$ such that $|x - z| < 1/n$ but $|1/x - 1/z| \geq \frac{1}{2}$, thus showing that $f(x) = 1/x$ is not uniformly continuous on $(0, 1)$. Show that f is uniformly continuous on $(1, 2)$.

6. Give an equivalent definition of uniform continuity using
   a) neighborhoods,
   b) sequences.

7. Show that if f is uniformly continuous on D then f is continuous on D.

8. Show that the sequence (z_{n_k}) in the proof of theorem 4.10 does converge to x_0.

9. Prove a generalized version of theorem 4.10 where the domain of the continuous function is a closed and bounded subset of $\mathbb{R}$ instead of a closed interval.

10. Fill in a detail of the proof of theorem 4.11 by showing why $\lim f(a_{n_k}) = f(\lim a_{n_k}) = f(x_0)$ and $f(a_{n_k}) \geq n_k$ is a contradiction.

11. Show that if $f:[a, b] \to \mathbb{R}$ is continuous on $[a, b]$ then f is bounded below and attains its minimum value.

12. Give an example of a function f which is continuous on its domain D such that:
    a) f is bounded above but does not attain its maximum.
    b) f is unbounded.
    c) f is bounded, attains its maximum but not its minimum.

13. Complete the proof of theorem 4.12 by dealing with the case $f(a) \geq f(b)$.

14. Prove the following generalization of theorem 4.12: Let $f:D \to \mathbb{R}$ be continuous on D. If D is an interval, then $f(D)$ is also an interval.

15. Give an alternative proof of the fact that a function continuous on a closed interval is bounded above by using the following idea, filling in details as needed:

    Let $f:[a, b] \to \mathbb{R}$ be continuous on $[a, b]$. Then we know that f is uniformly continuous on $[a, b]$ so there exists a $\delta_0 > 0$ such that if $x, z \in [a, b]$ with $|x - z| < \delta_0$ then $|f(x) - f(z)| < 1$. Thus, for all $x \in [a, b], f(x) \leq f(a) + (b - a)/\delta_0$.

16. Prove the following fact which was used implicitly in the proof of theorem 4.12: let $f:D \to \mathbb{R}$ be continuous on D, let (a_n) be a sequence in D such that for all $n \in \mathbb{N}, f(a_n) < c$. Suppose that $\lim a_n = L \in D$. Then $f(L) \leq c$. Give an example to show that the result need not hold if f is not continuous at L.

17. ***Believe It or Not***: Conjecture: Let $f:D \to \mathbb{R}$ be continuous at $x_0 \in D$. If $f(x_0) > 0$ then there exists $\delta > 0$ such that for all $x \in N_\delta(x_0) \cap D, f(x) > 0$.

    "Proof": Suppose that $f:D \to \mathbb{R}$ is continuous on D and that $x_0 \in D$ with $f(x_0) > 0$. Since f is continuous at x_0 there exists $\delta > 0$ such that if $x \in N_\delta(x_0) \cap D$ then $|f(x) - f(x_0)| < f(x_0)/2$. But this implies that $f(x) > f(x_0)/2 > 0$ for all $x \in N_\delta(x_0) \cap D$, as desired. $\square$

"Counterexample": Define $f: \mathbb{R} \to \mathbb{R}$ by $f(x) = |x|/x$ if $x \neq 0$ and $f(0) = 1$. Then $f(0) = 1 > 0$ but every neighborhood of 0 contains elements $x < 0$ for which $f(x) = -1 < 0$.

18. *Believe It or Not*: Conjecture: Let $f: D \to \mathbb{R}$ be continuous at $x_0 \in D$. Then there exists $\delta > 0$ such that f is continuous on $N_\delta(x_0) \cap D$.

"Proof": Suppose that f is continuous at x_0 and that for all $\delta > 0$ there exists an $x \in N_\delta(x_0) \cap D$ such that f is not continuous at x. We construct a sequence (a_n) as follows: for each $n \in \mathbb{N}$ let $a_n \in N_{1/n}(x_0) \cap D$ with f not continuous at a_n. Then $\lim a_n = x_0$ so $\lim f(a_n) = f(x_0)$ but f is discontinuous at each a_n, a contradiction. $\square$

"Counterexample": Let $f: \mathbb{R} \to \mathbb{R}$ be defined by

$$f(x) = \begin{cases} x & \text{if } x \text{ is rational,} \\ 0 & \text{if } x \text{ is irrational.} \end{cases}$$

Then clearly f is continuous at 0 but nowhere else.

CHAPTER
5

GROUPS

5.1 INTRODUCTION

If you were asked what algebra is about, after some thought you might give
an answer like, "Algebra is about solving equations." This would be a good
answer, for one of the main objectives of algebra is the solution of equations.
Most of the techniques you were taught in algebra, such as factoring and
methods of simplification, were used to help you solve equations. One thing
you might have noticed is that whether equations have solutions (or certain
expressions can be factored into linear factors, which is the same thing)
depends upon the number system you are using. For example, the equation
$x + 1 = 0$ has no solutions in $\mathbb{N}$ but does have a solution in $\mathbb{Z}$. Most of high
school algebra takes place in $\mathbb{R}$ so that perhaps you were told that $x^2 + 1$
cannot be factored. While this is true when one is restricted to $\mathbb{R}$, there is a
factorization, $x^2 + 1 = (x + i)(x - i)$, if one is willing to include elements
of $\mathbb{C}$, the set of complex numbers. This brings up the question: how much of
algebraic "behavior" is due to the particular operations involved (in our case
addition and multiplication) and how much is due to the particular objects
upon which these operations are performed? One of the ways to investigate
questions of this sort is to abstract the "essential" features of the operations
through a suitable selection of axioms and see what can be shown. In this
sort of context-free environment we can focus on general behavior and not be
distracted by particulars. In many cases we don't know what the operations
are nor do we know what sort of objects make up the underlying set. While
this may seem strange at first (and perhaps somewhat disconcerting) it turns

147

out to be a very useful situation. By only specifying how a system "acts," then anything we can prove about it can be applied to any other system which acts the "same." This leads to one of the central themes in abstract algebra, that of two structures being *isomorphic* or having the same form. We will examine this and some other ideas in the short introduction to abstract algebra which follows.

5.2 GROUPS

The general situation we will investigate is one in which we have a non-empty set of objects with one binary operation which satisfies certain axioms. We begin with

---

Definition 5.1: If G is a non-empty set and $\star$ is a binary operation on G such that

a) $\star$ is associative,

b) $\star$ has an identity (which we often denote by e), and

c) every element of G has an inverse with respect to $\star$,

then we say, $(G, \star)$ is a *group*. If there is no danger of confusion (and even sometimes when there is), by a slight abuse of notation, we will often refer to the group as G. If $\star$ is commutative, we say $(G, \star)$ is an *abelian* group.

When we write the operation as multiplication (using $\cdot$ or juxtaposition) we will denote the inverse of a by a^{-1}; if we write the operation as $+$ we will denote the inverse of a by $-a$. It is important to remember that "built in" to the definition of a binary operation is the notion of closure; if $\star$ is a binary operation on G and $a, b \in G$ then $a \star b \in G$.

Here are some examples of this idea, some familiar and some perhaps not so familiar:

a) $(\mathbb{Z}, +)$ is an abelian group with identity 0.

b) $(\mathbb{Z}, \cdot)$ is not a group, for although the operation is associative and there is an identity (1), not every element has an inverse (which do?).

c) $(\mathbb{Q}, +)$ is an abelian group with identity 0.

d) $(\mathbb{Q}, \cdot)$ is not a group because 0 has no inverse. The same is true if we substitute $\mathbb{R}$ for $\mathbb{Q}$.

e) $(\mathbb{Q} - \{0\}, \cdot)$ is an abelian group with identity 1. The same is true if we substitute $\mathbb{R}$ for $\mathbb{Q}$.

f) Let A be a non-empty set and let $S(A) = \{f : f : A \to A \text{ is a bijection}\}$. Then $(S(A), \circ)$ (where $\circ$ is functional composition) is a group, usually non-abelian, with identity I_A. The elements of this group are called *permutations* of A and this group is called the *permutation group* on A.

g) Let $G = \{f : f : \mathbb{R} \to \mathbb{R}, f(0) = 0\}$ and for $f, g \in G, x \in \mathbb{R}$, define $(f + g)(x) = f(x) + g(x)$. Then $(G, +)$ is an abelian group with identity $z : \mathbb{R} \to \mathbb{R}$ defined by $z(x) = 0, \forall x \in \mathbb{R}$.

h) Let $n \in \mathbb{N}$. Let $\mathbb{Z}(n) = \{0, 1, \ldots, n - 1\}$. For $p, q \in \mathbb{Z}(n)$ we define

$$p \oplus_n q = r \text{ where } r \in \mathbb{Z}(n) \text{ and } r \equiv p + q \pmod{n}.$$

Then $(\mathbb{Z}(n), \oplus_n)$ is an abelian group with identity 0.

i) Let A be a non-empty set and let $G = \mathbb{P}(A)$. For $B, C \in G$ we define

$$B + C = (B - C) \cup (C - B).$$

Then $(G, +)$ is an abelian group with identity $\emptyset$ and every element is its own inverse.

j) Let $G = \{a, b, c, d\}$ with the group operation given in the table below, where, for example, the entry in row b and column c is bc; hence, $bc = d$. These tables are called *Cayley tables* after the English mathematician of the same name. That this is an associative operation (it is) is not obvious, but we should be able to see it is a commutative operation with identity a.

	a	b	c	d
a	a	b	c	d
b	b	a	d	c
c	c	d	a	b
d	d	c	b	a

k) Let $G = \{a, b, c, d\}$ with the group operation given in the Cayley table below:

	a	b	c	d
a	a	b	c	d
b	b	c	d	a
c	c	d	a	b
d	d	a	b	c

Again, G is an abelian group with identity a.

l) Let $G_6 = \{a, b, c, d, e, f\}$ (we give this group a special name for we will be referring to it again) with the group operation given in the Cayley table below.

	e	a	b	c	d	f
e	e	a	b	c	d	f
a	a	e	c	b	f	d
b	b	d	e	f	a	c
c	c	f	a	d	e	b
d	d	b	f	e	c	a
f	f	c	d	a	b	e

$$G_6$$

In this case, G is a non-abelian group with identity e.

These last three examples are instances of the situation described earlier in which we don't know what the elements are nor do we know what the operation is.

Recalling some of our results about binary operations from chapter 2, we see that in a group inverses and identities are unique, and that $(ab)^{-1} = b^{-1}a^{-1}$ for all elements a, b in the group. Some other results which hold for all groups are

---

Theorem 5.1: Let $(G, \cdot)$ be a group with identity e.

a) $\forall a, b, c \in G, ab = ac$ implies $b = c$ (this is called the *cancellation law*).

b) $\forall a, b \in G, \exists x \in G \ni ax = b$.

c) Let $x \in G$. If $xy = e$ then $x = y^{-1}$ and $y = x^{-1}$.

d) Let $x \in G$. If there exists a $y \in G$ such that $xy = y$ then $x = e$.

Proof: a) Let $a, b, c \in G$ with $ab = ac$. Then

$$b = eb = (a^{-1}a)b = a^{-1}(ab) = a^{-1}(ac) = (a^{-1}a)c = ec = c.$$

b) Let $a, b \in G$. Then since $a(a^{-1}b) = (a^{-1}a)b = b$, the equation $ax = b$ has the (unique) solution $x = a^{-1}b$.

c) Let $x, y \in G$ with $xy = e$. Then $(xy)y^{-1} = ey^{-1}$ or $x = y^{-1}$.
Also, $x^{-1}(xy) = x^{-1}e$ so $y = x^{-1}$.

d) Let $x, y \in G$ with $xy = y$. Then $(xy)y^{-1} = yy^{-1}$ so $x = e$. $\square$

One of the ways we can get some insight into groups is to examine those subsets which form groups:

Definition 5.2: Let $(G, \cdot)$ be a group and $H \subseteq G$. If, with $\cdot|_{H \times H}$ as the operation, $(H, \cdot)$ is a group, we say H is a *subgroup of* G.

In this connection, recall (from chapter 2) that if $\cdot|_{H \times H}$ is a binary operation on H we say that H is *closed* with respect to $\cdot$.

Here are some examples of subgroups:

a) For any group G, $\{e\}$ and G are subgroups.

b) In $(\mathbb{Z}, +)$, $n\mathbb{Z} = \{nk : k \in \mathbb{Z}\}$ is a subgroup for each $n \in \mathbb{Z}$.

c) For any group G, $Z(G) = \{g \in G : \forall x \in G, gx = xg\}$ is a subgroup of G. $Z(G)$ is called the *center of* G.

d) In the group of functions $(G, +)$ given above in example g), $H_\alpha = \{f \in G : f(\alpha) = 0\}$ is a subgroup of G for each $\alpha \in \mathbb{R}$.

e) For the group G_6 given as example l) above, the only subgroups are

$$G_6, \{e\}, \{e, a\}, \{e, b\}, \{e, f\}, \{e, c, d\}.$$

To prove that a subset H of a group $(G, \cdot)$ is a subgroup, we could verify that $(H, \cdot)$ satisfies all the axioms (this would include showing that H is closed with respect to $\cdot$); however, it turns out that there is an easier way:

Theorem 5.2: Let $(G, \cdot)$ be a group and H a non-empty subset of G. Then $(H, \cdot)$ is a subgroup of $(G, \cdot)$ if and only if

$$\forall a \in H, a^{-1} \in H \text{ and } \forall a, b \in H, ab \in H.$$

Proof: Clearly, if H is a subgroup of G then these two conditions must be satisfied. Conversely, if H is a non-empty subset of G which satisfies these two conditions then H is closed with respect to $\cdot$ and so $\cdot$ is an associative binary operation on H. Also, it is clear that each element of H has an inverse in H, so all that remains to be shown is that H has an identity. Since

H is non-empty, let $a \in H$. Then $a^{-1} \in H$ and as H is closed with respect to $\cdot$ we have $aa^{-1} = e \in H$. Thus H is a subgroup of G. $\square$

We can create some new (and very interesting) objects with subgroups:

Definition 5.3: Let G be a group and H a subgroup of G. If $x \in G$, $xH = \{xh : h \in H\}$ is called a *left coset of* H and $Hx = \{hx : h \in H\}$ is called a *right coset of* H.

There are a few basic facts about cosets which we will need in our future work; we will prove them for left cosets, but of course similar results hold for right cosets:

Theorem 5.3: Let G be a group and H a subgroup of G. Then:

a) $\forall a \in G, a \in aH$.
b) $\forall a, b \in G, a \in bH$ if and only if $aH = bH$.
c) $\forall a, b \in G, aH \subseteq bH$ implies $aH = bH$.
d) $\forall a, b \in G, aH = bH$ if and only if $aH \cap bH \neq \emptyset$.
e) $\{aH : a \in G\}$ is a partition of G.
f) $\forall a \in G, aH = Ha$ if and only if $a^{-1}Ha = H$.
g) $\forall a, b \in G, aH = bH$ if and only if $a^{-1}b \in H$.

Proof: Let G be a group and H a subgroup of G.

a) Suppose $a \in G$. Since H is a subgroup, $e \in H$ so $ae = a \in aH$.

b) Suppose that $a, b \in G$ with $a \in bH$. Then there exists $h_1 \in H$ such that $a = bh_1$. Let $ah_2 \in aH$. Then $ah_2 = bh_1h_2 \in bH$ so $aH \subseteq bH$. Now let $bh_3 \in bH$. But $bh_3 = ah_1^{-1}h_3 \in aH$ so we have $bH \subseteq aH$ and hence $aH = bH$. For the converse, suppose that $aH = bH$. Then $a = ae \in aH = bH$.

c) Suppose that $a, b \in G$ with $aH \subseteq bH$. Let $bh_1 \in bH$. Since $a \in aH \subseteq bH$, there exists $h_2 \in H$ such that $a = bh_2$. Thus $bh_1 = ah_2^{-1}h_1 \in aH$ so $bH \subseteq aH$ and we have $aH = bH$.

d) Clearly, if $aH = bH$ then $aH \cap bH = aH \neq \emptyset$. To prove the converse, suppose that $a, b \in G$ with $aH \cap bH \neq \emptyset$. Let $g \in aH \cap bH$. Thus there exist $h_1, h_2 \in H$ such that $g = ah_1 = bh_2$. Hence $a = bh_2h_1^{-1} \in bH$ so by b) we have $aH = bH$.

e) This follows immediately from a) and d).

f) Left as an exercise.

g) Suppose that $a, b \in G$ with $aH = bH$. Then we know $a \in bH$ so there exists $h_1 \in H$ such that $a = bh_1$. Thus

$$a^{-1}b = (bh_1)^{-1}b = h_1^{-1}b^{-1}b = h_1^{-1} \in H.$$

For the converse, suppose that $a^{-1}b \in H$. Then there exists $h \in H$ such that $a^{-1}b = h$ which implies $b = ah \in aH$ so by part b) $aH = bH$. $\qquad\qquad\square$

It is "natural" (to mathematicians, anyway), now that we have some new objects (cosets) in the context of a group, to try to extend the group operation to these new objects. If G is a group and H is a subgroup of G, we might start by defining $(aH)(bH)$ to be $(ab)H$. Let's look at an example to see what is involved. Let G_6 be the group given in example m) (for convenience, its Cayley table is given below).

	e	a	b	c	d	f
e	e	a	b	c	d	f
a	a	e	c	b	f	d
b	b	d	e	f	a	c
c	c	f	a	d	e	b
d	d	b	f	e	c	a
f	f	c	d	a	b	e

$$G_6$$

Let $H = \{e, a\}$. Then (you should check all these computations) $bH = \{b, d\}$, $cH = \{c, f\}$ and since $bc = f$ we would like to define $bHcH = fH = \{f, c\}$. This seems all right, but (and here is where the trouble begins) since $bH = dH$ and $cH = fH$ we must have $bHcH = bcH = dHfH = dfH$ if our binary operation is to be well-defined (that is, if it is to be a function). However, $df = a$ and $aH = H$ so in fact, we do not have a binary operation. Rather than give up our idea completely, let's see what we need in order to have our operation well-defined. If H is a subgroup of G and $a, b, c, d \in G$ with $aH = bH$, $cH = dH$, then we need $acH = bdH$. But this will be the case if and only if $(ac)^{-1}bd \in H$ or $c^{-1}a^{-1}bd \in H$. But $aH = bH$ so $a^{-1}b \in H$. Also we have $c^{-1}d \in H$

so there exist $h_1, h_2 \in H$ such that $a^{-1}b = h_1$ and $c^{-1}d = h_2$. Let's see what we now have:

$$c^{-1}a^{-1}bd = c^{-1}h_1d = h_2d^{-1}h_1d = c^{-1}h_1ch_2.$$

This is getting close, for if c or d commuted with elements of H we would have the desired result. Actually we do not need quite this much; what we do need is given in the following definition:

Definition 5.4: Let G be a group and H a subgroup of G. We say H is a *normal* subgroup of G if and only if $\forall a \in G, aH = Ha$.

Thus, if H is a normal subgroup of G and $a \in G, h \in H$, then we have $ah \in aH = Ha$ so there is an element of H, say h_1, such that $ah = h_1a$. This is not commutativity, but it is all that we needed above to be able to finish off: if H is normal then there is an $h_3 \in H$ such that $c^{-1}h_1 = h_3c^{-1}$ so

$$c^{-1}h_1ch_2 = h_3c^{-1}ch_2 = h_3h_2 \in H.$$

Hence our operation is well-defined on cosets of a normal subgroup. In fact, things are better than we might have hoped for:

Theorem 5.4: Let $(G, \cdot)$ be a group and $(H, \cdot)$ a normal subgroup of G. Then $G/H = \{aH : a \in G\}$ with the operation defined by $\forall a, b \in G, (aH) \cdot (bH) = (a \cdot b)H$ is a group.

Proof: Suppose that H is a normal subgroup of G. Our previous discussion has shown that the operation we have defined on G/H is in fact a binary operation, so we must show that it is associative, find an identity and show that every element (coset) has an inverse. To see that the operation is associative, let $a, b, c \in G$. Then

$$(aH)(bHcH) = (aH)(bcH) = a(bc)H$$
$$= (ab)cH = (ab)HcH = (aHbH)(cH),$$

as required. Also, if $a \in G$, then since $eH = H, aHH = aHeH = aeH = aH$ and $HaH = eHaH = eaH = aH$ so H is the identity. As we might expect, if $a \in G$, $a^{-1}HaH = a^{-1}aH = H = aa^{-1}H = aHa^{-1}H$ so $(aH)^{-1} = a^{-1}H$ and every element of G/H has an inverse. Thus the set of cosets of a normal subgroup is a group with the induced group operation. This is sometimes called the *factor* or *quotient* group of G

modulo H. Note that we only used the assumption that H was normal to ensure that the operation was well-defined. □

Another way of looking at factor groups is to recall that partitions and equivalence relations are equivalent ideas. Thus we might start with an equivalence relation on a group G, say $\sim$, and attempt to induce an operation on the set of equivalence classes. For this to be a binary operation the equivalence relation must be "compatible" with the operation; i.e., if $a, b, c, d \in G$ with $a \sim b$ and $c \sim d$ then we need $ac \sim bd$. It turns out (details left as exercise 20) that the partition induced by such an equivalence relation is just a set of cosets of a normal subgroup, so we don't get anything new.

Exercises 5.2

1. Exactly when is the group mentioned in group example f) non-abelian?

2. If we modify example g) by letting $G = \{f : f : \mathbb{R} \to \mathbb{R}, f(0) = 1\}$ and use the same operation, do we get a group?

3. In the Cayley table for a group, each element appears exactly once in each row and each column. Explain why this is the case.

4. Show that if a, b are in a group G then the equation $ax = b$ has exactly one solution.

5. Suppose that G is a group with the property that for all $a, b \in G, a^2b^2 = (ab)^2$. Show that G is abelian.

6. Give an example of a non-empty subset of a group which is closed with respect to the operation but is not a subgroup.

7. Let $(G, \cdot)$ be a group and H a non-empty subset of G. Show that $(H, \cdot)$ is a subgroup of $(G, \cdot)$ if and only if $\forall a, b \in H, ab^{-1} \in H$.

8. Let G be a group and H, K subgroups of G. Show that
   a) $H \cap K$ is a subgroup of G.
   b) $H \cup K$ is a subgroup of G if and only if $H \subseteq K$ or $K \subseteq H$.

9. State and prove a result for right cosets which corresponds to part g) of theorem 5.3.

10. Suppose that we had defined cosets using arbitrary non-empty subsets H of G rather than subgroups. Which of the properties of theorem 5.3 would still hold? Prove any that do and give examples to show that the others do not necessarily hold.

11. Prove part f) of theorem 5.3.

12. Let G be a group and H a subgroup of G. We define a relation on G by $a \equiv b \pmod{H}$ if and only if $a^{-1}b \in H$. Show that this relation

is an equivalence relation on G and that the partition it induces is the same as the partition by left cosets of H.

13. (Continuation of exercise 12) We now have two similar-looking notations for two equivalence relations; $a \equiv b \pmod{H}$ where H is a subgroup of a group G and congruence modulo an integer n on $\mathbb{Z}$, $a \equiv b \pmod{n}$. Show that if we take G to be $(\mathbb{Z}, +)$ and H to be $n\mathbb{Z}$ that these two are in fact the same. Thus congruence modulo a subgroup is a generalization of congruence modulo an integer.

14. In the group G_6 given as example 1), list all the right and left cosets of the subgroups $H = \{e, a\}$ and $K = \{e, c, d\}$. From this listing, determine the normality of these subgroups.

15. Let G be a group. Show that $Z(G)$ is a normal subgroup of G.

16. Let G be a group and suppose that $a \in G$ is the only element aside from e such that $a^2 = e$. Show that $a \in Z(G)$.

17. Let G be a group with H, K normal subgroups of G such that $H \cap K = \{e\}$. Show that for all $h \in H, k \in K, hk = kh$.

18. Let G be a group and H a subgroup of G. For $a, b \in G$ we define a function $f: aH \to bH$ by $f(ah) = bh$. Show that f is a bijection. Thus, if H is finite, then all the left cosets of H have the same number of elements. Use this and the fact that the left cosets form a partition of G to prove that the number of elements in H divides the number of elements in a finite group G.

19. Let G be a group and H a subgroup of G. Show that the following are equivalent:
    a) H is a normal subgroup of G.
    b) $\forall a \in G, a^{-1}Ha = H$.
    c) $\forall a \in G, \exists b \in G \ni aH = Hb$.
    d) $\forall a, b \in G, ab \in H$ implies $ba \in H$.

20. Let $(G, \cdot)$ be a group and $\sim$ an equivalence relation on G which is compatible with $\cdot$. Show that:
    a) $[e]_\sim$ is a subgroup of G.
    b) $[e]_\sim$ is a normal subgroup of G.
    c) $\forall a \in G, [a]_\sim = a[e]_\sim$.

21. *Believe It or Not:* Conjecture: Let G be a group and H a normal subgroup of G. Then G/H is abelian if and only if G is abelian.

    "Proof": If G is abelian it is clear that G/H is abelian, for if $a, b \in G$ then $aHbH = abH = baH = bHaH$. To prove the converse, suppose that G/H is abelian and let $a, b \in G$. Thus $aHbH = bHaH$, so we have $abH = baH$. Therefore $ab = ba$ and G is abelian. $\square$

    "Counterexample": Let G be any non-abelian group. Then $G/G = \{G\}$ is abelian but G is not.

22. **_Believe It or Not:_** Conjecture: Let H, K be subgroups of a group G. Then $HK = \{hk : h \in H, k \in K\}$ is a subgroup of G.

"Proof": Since $ee = e \in HK$, $HK \neq \varnothing$. Suppose that $a, b \in HK$. Then there exist $h_1, h_2 \in H, k_1, k_2 \in K$ such that $a = h_1 k_1, b = h_2 k_2$. Thus $ab = h_1 k_1 h_2 k_2 = h_1 (h_2^{-1} k_1^{-1})^{-1} k_2 = h_1 h_2 k_1 k_2 \in HK$. Also, $a^{-1} = (h_1 k_1)^{-1} = h_1^{-1} k_1^{-1} \in HK$ so by theorem 5.2 HK is a subgroup of G. □

"Counterexample": Let G_6 be the group given in example 1). If we let $H = \{e, a\}, K = \{e, b\}$ then $HK = \{e, b, a, c\}$ which is not a subgroup of G_6.

5.3   GROUPS AND FUNCTIONS

An important activity in algebra is determining if two algebraic structures are the same. For example, consider the two groups with four elements shown in the Cayley tables below:

	e	a	b	c			e	a	b	c
e	e	a	b	c		e	e	a	b	c
a	a	e	c	b		a	a	b	c	e
b	b	c	e	a		b	b	c	e	a
c	c	b	a	e		c	c	e	a	b

$$G_1 \qquad\qquad\qquad G_2$$

Clearly, e is the identity of each group. After a little inspection, we see that these two groups are not the "same" since in G_1 every element satisfies $x^2 = e$, that is, every element is its own inverse, while in G_2 only two elements (e, b) satisfy this equation. What if we had not noticed something so obviously different? How could we have shown that two groups are the same? Functions to the rescue!

Definition 5.5:   Let $(G, \cdot), (H, \star)$ be groups and $f : G \to H$. We say that f is a _homomorphism_ if

$$\forall x, y \in G, f(x \cdot y) = f(x) \star f(y).$$

If f is also a bijection we say f is an _isomorphism_ and that G is _isomorphic_ to H (denoted by $G \cong H$). If $f : G \to G$ is an isomorphism, f is called an _automorphism_.

If G, H are any groups, then $f: G \to H$ defined by $\forall x \in G, f(x) = e$ is a homomorphism. Also, I_G is an automorphism of G. If two groups are isomorphic then algebraically they have the same form, for not only is there a one-to-one correspondence between the elements of the sets but also between products of pairs of corresponding elements. Thus, whatever can be said (algebraically) of one group is also true of the other, even though the elements of one group may be functions and those of the other numbers (or whatever) and the operations may likewise be very different.

There are a few basic facts about homomorphisms which we need to record for future use:

Theorem 5.5: Let $(G, \cdot), (H, \star)$ be groups and $f: G \to H$ a homomorphism.

a) Let e be the identity of G. Then $f(e)$ is the identity of H.
b) $\forall x \in G, (f(x))^{-1} = f(x^{-1})$.
c) $Im(f)$ is a subgroup of H.
d) Let e' be the identity of H. Then $f^{-1}(\{e'\}) = \{x \in G : f(x) = e'\}$ is a normal subgroup of G. This set is called the *kernel* of f, denoted by $ker\, f$.

Proof: Let $(G, \cdot), (H, \star)$ be groups, $f: G \to H$ a homomorphism and e, e' the identities of G and H, respectively. For a), we observe that $f(e) \star f(e) = f(ee) = f(e)$, so $f(e) = e'$. To prove b), let $x \in G$. Then

$$e' = f(e) = f(xx^{-1}) = f(x) \star f(x^{-1}),$$

so $f(x^{-1}) = (f(x))^{-1}$. To show that $Im(f)$ is a subgroup of H, we first observe that $Im(f) \neq \varnothing$, since $f(e) \in Im(f)$. Now suppose that $a, b \in Im(f)$. Then there exist $x, y \in G \ni f(x) = a, f(y) = b$. But $f(xy) = f(x) \star f(y) = a \star b$ so $Im(f)$ is closed with respect to $\star$. Also $f(x^{-1}) = (f(x))^{-1} = a^{-1}$. To show that $ker\, f$ is a subgroup of G, we first note that $e \in ker\, f$ since $f(e) = e'$. Suppose that $a, b \in kerf$. Then

$$f(ab) = f(a) \star f(b) = e' \star e' = e'$$

so $ab \in ker\, f$. Also

$$f(a^{-1}) = (f(a))^{-1} = (e')^{-1} = e'$$

so $a^{-1} \in ker\, f$ and by theorem 5.2, $ker\, f$ is a subgroup of G. Now we show that $ker\, f$ is a normal subgroup of G. Let $a \in G$ and $k \in ker\, f$. Then

$$f(a^{-1}ka) = (f(a))^{-1} \star f(k) \star f(a) = (f(a))^{-1} \star f(a) = e'$$

so $a^{-1}ka \in ker\ f$. But then $ka \in a(ker\ f)$ so $(ker\ f)a \subseteq a(ker\ f)$. A similar argument shows that $a(ker\ f) \subseteq (ker\ f)a$ so $ker\ f$ is a normal subgroup of G. □

Thus we see that kernels of homomorphisms are normal subgroups. A surprising fact is that every normal subgroup is the kernel of a homomorphism so that in a certain sense these ideas are equivalent.

Theorem 5.6: Let G be a group and N a normal subgroup of G. We define $\alpha: G \to G/N$ by $\alpha(x) = xN, \forall x \in G$. Then α is a homomorphism with $ker\ \alpha = N$.

Proof: Let N be a normal subgroup of G and define $\alpha: G \to G/N$ by $\alpha(x) = xN$ for all $x \in G$. Suppose $x, y \in G$. Then $\alpha(xy) = xyN = xNyN = \alpha(x)\alpha(y)$ so α is a homomorphism (you should be able to supply the justification for each of these equalities). Since N is the identity of G/N, we must show that $\alpha(N) = N$. But this is immediate since $\alpha(a) = aN = N$ if and only if $a \in N$. □

This sets the stage for what is known as the *fundamental theorem of group homomorphisms:*

Theorem 5.7: Let G, H be groups with $f: G \to H$ a homomorphism with $ker\ f = K$. Then $G/K \cong Im(f)$.

Proof: It may be helpful to consider the (familiar-looking) diagram below:

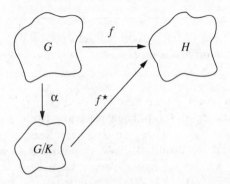

If we define $f^{\star}: G/K \to H$ by $f^{\star}(aK) = f(a)$ for all $aK \in G/K$ and observe that (from exercise 3 in this section) the cosets of K are just the subsets of G on which f is constant, we can use the results of theorem 2.21 to see that $f^{\star}$ is in fact a function (that is, well-defined, always a question which comes up when we define a function on equivalence classes). We also know that $f^{\star}$ is a bijection onto $Im(f)$, so all that remains to be proven is that $f^{\star}$ is a homomorphism. But if $aK, bK \in G/K$ then

$$f^{\star}(aK\,bK) = f^{\star}(abK) = f(ab) = f(a)f(b) = f^{\star}(aK)f^{\star}(bK),$$

which completes the proof. $\qquad\qquad\qquad\qquad\qquad\qquad\qquad\qquad$ □

As an application of this theorem consider $(\mathbb{Z}, +)$, $(\mathbb{Z}(5), +)$ and $f: \mathbb{Z} \to \mathbb{Z}(5)$ defined by $f(n) = r$ where $0 \le r < 5$ and $n \equiv r \pmod 5$. In this case, $ker f = \{\ldots, -5, 0, 5, 10, \ldots\}$ (since 0 is the identity of $\mathbb{Z}(5)$) and $\mathbb{Z}/ker f = \mathbb{Z}_5$ (the set of equivalence classes of congruence modulo 5). Since f is surjection, by theorem 5.7 we have $\mathbb{Z}_5 \cong \mathbb{Z}(5)$.

For another application of this theorem, we need to develop some information about automorphisms. First, if G is a group, we will denote the set of automorphisms of G by $A(G)$. Thus,

$$A(G) = \{f : f: G \to G \text{ is an automorphism}\}.$$

It turns out (details left as an exercise) that $(A(G), \circ)$ is a group (I_G is the identity). In general it is difficult to find all the automorphisms of a given group, but there is a certain type which is easy to find:

Definition 5.6: Let G be a group and $g \in G$. Then $f_g: G \to G$ defined by $f_g(x) = g^{-1}xg, \forall x \in G$ is called an *inner automorphism* of G.

Note that $f_e = I_G$. That functions defined in this way are in fact automorphisms is left as an exercise. The set of inner automorphisms of a group is a subgroup of the automorphism group:

Theorem 5.8: Let G be a group and let $I(G) = \{f_g : g \in G\}$, the set of inner automorphisms on G. Then $I(G)$ is a subgroup of $A(G)$.

Proof: Let G be a group. As we have observed earlier, $f_e = I_G$ so $I(G) \ne \varnothing$. Suppose that $f_g, f_h \in I(G)$. Then we know (by exercise 9 at the end of this section) that $f_g f_h = f_{hg}$ so $I(G)$ is closed with respect to the operation (functional composition). Now suppose that $f_g \in I(G)$. Let $x \in G$. Then $(f_g \circ f_{g^{-1}})(x) = f_g(gxg^{-1}) = g^{-1}gxg^{-1}g = x$ so $f_g f_{g^{-1}} = I_G$ or $(f_g)^{-1} = f_{g^{-1}} \in I(G)$. Thus $I(G)$ is a subgroup of $A(G)$. $\qquad$ □

Now we are ready for our other application of theorem 5.7:

Theorem 5.9: Let G be a group. Let $Z(G)$ be the center of G and $I(G)$ the group of inner automorphisms of G. Then $G/Z(G) \cong I(G)$.

Proof: We first observe that at least this makes sense from a quantitative point of view, for the "more abelian" G is, that is, the more elements in $Z(G)$, the fewer distinct inner automorphisms G has. To apply theorem 5.7 we need to find a homomorphism from G onto $I(G)$ whose kernel is $Z(G)$. An "almost natural" candidate (probably our second choice) is $h: G \to I(G)$ defined by $h(x) = f_{x^{-1}}$ for all $x \in G$. Clearly, h is a surjection. If $x, y \in G$ then

$$h(xy) = f_{(xy)^{-1}} = f_{y^{-1}x^{-1}} = f_{x^{-1}}f_{y^{-1}} = h(x)h(y),$$

so h is a homomorphism. Next we show $ker\ h = \{x \in G : h(x) = I_G\}$ $= Z(G)$. If $z \in Z(G)$, then $h(z) = f_{z^{-1}}$ and $\forall x \in G, f_{z^{-1}}(x) = zxz^{-1}$ $= xzz^{-1} = x$, so $h(z) = I_G$ and $z \in ker\ h$. That $ker\ h \subseteq Z(G)$ is left as exercise 11. Thus, by theorem 5.7 we have $G/ker\ h = G/Z(G)$ $\cong I(G)$. $\qquad\square$

This completes our brief introduction to algebra. We have barely scratched the surface of group theory, having left out many interesting and important aspects, to say nothing of algebraic structures with more than one binary operation. One thing which we hope has come across, though, is the important role played by functions; in particular the importance of homomorphisms.

Exercises 5.3

1. Let G, H be groups and $f: G \to H$ defined by $g(x) = e, \forall x \in G$. Show that f is a homomorphism.

2. Let S be a non-empty set of groups. Show that $\cong$ (isomorphism) is an equivalence relation on S.

3. Let G, H be groups and $f: G \to H$ be a homomorphism with $K = ker f$. Show that $aK = \{x \in G : f(x) = f(a)\}$. This shows that the equivalence relation which gives rise to the partition of G into cosets of $ker f$ is just the equivalence relation R where xRy if and only if $f(x) = f(y)$ which we studied in chapter 2.

4. Which of the two groups of four elements shown at the beginning of this section is isomorphic to $(\mathbb{Z}(4), +)$?

5. Let G be a group. Show that $(A(G), \circ)$ is a group.

6. Let G be a group and $g \in G$. Show that $f_g: G \to G$ defined by $f_g(x) = g^{-1}xg, \forall x \in G$ is an automorphism of G.

7. If G is an abelian group, how many distinct inner automorphisms does G have? Give reasons for your answer.

8. Let G be an abelian group. Show that $f: G \to G$ defined by $f(x) = x^{-1}, \forall x \in G$ is an automorphism. Also show that if G is not abelian then f is not an automorphism. What can be said about f defined by $f(x) = x^2$?

9. Let G be a group with $g_1, g_2 \in G$. Show that $f_{g_1} \circ f_{g_2} = f_{g_2 g_1}$.

10. Prove or give a counterexample: if G is a group then $I(G)$ is a normal subgroup of $A(G)$.

11. Complete the proof of theorem 5.8 by showing that $ker\ h \subseteq Z(G)$.

12. List the elements of $Z(G_6)$ and then find the cosets of $Z(G_6)$. Find the distinct inner automorphisms of G_6. Is there any connection between the cosets of $Z(G_6)$ and the inner automorphisms?

13. Let G be a group with f_x, f_y inner automorphisms of G. Show that $f_x = f_y$ if and only if $xy^{-1} \in Z(G)$.

14. Show that the "natural" choice for the homomorphism in the proof of theorem 5.9,

$$h: G \to I(G)$$

defined by

$$h(x) = f_x, \forall x \in G,$$

is in general not a homomorphism.

15. Let G, H be groups and $f: G \to H$ a homomorphism.
    a) Show that f is one-to-one if and only if $ker f = \{e\}$.
    b) If K is a normal subgroup of G, show that $f(K)$ is a normal subgroup of $Im(f)$.

16. Show that $(\mathbb{R}, +) \cong (\mathbb{R}_+, \cdot)$.

17. Show that if G is any group with 3 elements then $G \cong Z(3)$.

18. Let $H = \{f : f: \mathbb{R} \to \mathbb{R}, f$ is differentiable$\}$, $G = \{f : f: \mathbb{R} \to \mathbb{R}\}$ with the operation of addition (as defined in example g) in section 5.2). Show that $(H, +)$ is a subgroup of $(G, +)$. Let $\phi: H \to G$ be defined by $\phi(f) = f'$ (the derivative of f). Show that ϕ is a homomorphism. What is $ker\ \phi$?

19. Suppose that H is a subgroup of G and K is a normal subgroup of G. Show that:
    a) HK is a subgroup of G.
    b) $H \cap K$ is a normal subgroup of H.

c) K is a normal subgroup of HK.

d) $H/(H \cap K) \cong HK/K$.

20. Let G be a group. For each $g \in G$ we define $\alpha_g : G \to G$ by $\alpha_g(x) = xg$ for all $x \in G$.

a) Show that $\forall g \in G, \alpha_g$ is a bijection.

b) Recall that $(S(G), \circ)$ where

$$S(G) = \{f : f : G \to G \text{ is a bijection}\}$$

is a group. Show that $\phi : G \to S(G)$ defined by $\phi(g) = \alpha_g, \forall g \in G$ is a homomorphism.

c) Show that $G \cong Im(\phi)$ so that every group is isomorphic to a group of permutations.

21. *Believe It or Not:* Conjecture: Let G be a group with $f : G \to G$ a homomorphism. If $f(G)$ has exactly two elements then for all $g \in G, g^2 \in ker f$.

"Proof": Let G be a group and $f : G \to G$ a homomorphism with two elements in $f(G)$, say e and $a \neq e$. Let $x \in G$. If $f(x) = e$ then $e = f(x)f(x) = f(x^2)$ so $x^2 \in ker f$. If $f(x) = a$ then since $f(G)$ is a subgroup, $a^2 = e$ and we have $e = a^2 = f(x)f(x) = f(x^2)$. Thus $x^2 \in ker f$. $\qquad \square$

"Counterexample": Consider $(\mathbb{Z}(4), +)$ and $f : \mathbb{Z}(4) \to \mathbb{Z}(4)$ defined by $f(0) = f(2) = 0, f(1) = f(3) = 2$. Then f is a homomorphism and $f(\mathbb{Z}(4))$ has two elements but $1^2 = 1 \notin ker f = \{0, 2\}$.

22. *Believe It or Not:* Conjecture: If H is a subgroup of G and H has only two left cosets, then H is a normal subgroup of G.

"Proof": Let G be a group with H a subgroup which has only two left cosets, say H and aH. We define $f : G \to \mathbb{Z}(2)$ by $f(x) = 0$ if $x \in H$ and $f(x) = 1$ if $x \notin H$. We also observe that for $g_1, g_2 \in G, g_1 g_2 \in H$ if and only if g_1 and $g_2 \in H$ or g_1 and $g_2 \notin H$. Thus f is a homomorphism. But clearly we have $ker f = H$ so H is a normal subgroup of G. $\qquad \square$

"Counterexample": Consider G_6 with $H = \{e, c, d\}$. Then H has only two left cosets, H and aH, but since $ac = b \neq f = ca$ we don't have $gh = hg$ for all $g \in G, h \in H$ so H is not normal.

CHAPTER
6

CARDINALITY

6.1 INTRODUCTION

In this chapter we wish to make precise our idea of "the number of elements in a set" and then use this idea to distinguish between finite and infinite sets. We will call this concept *cardinality*. Cardinality is interesting in itself and its study will also give us a chance to apply some of the things we have learned about functions.

If we think about what we mean when we say "I have as many fingers as toes" we might notice that what we really mean is "there exists a bijection from my fingers to my toes." That is, there is a one-to-one correspondence between the set consisting of my fingers and the set consisting of my toes. One should not feel that one had been deprived as a child if this fact was not pointed out at one's mother's knee; while one-to-one correspondences are not rated "R," they generally are not explicitly mentioned in childhood conversations.

6.2 CARDINAL NUMBERS

To get us started, we define a relation which will embody the idea "set A has the same number of elements as set B."

Definition 6.1: Let A and B be sets. If there exists a one-to-one correspondence $f: A \rightarrow B$ we say that A is *equivalent* to B, denoted by $A \sim B$.

Intuitively, $A \sim B$ if and only if A and B have the same number of elements. For example, $\{1, 3, 4\} \sim \{3, 17, \pi\}$ while $\{1, 2, 3\} \not\sim \{1, 2, 4, 5\}$. Perhaps not so intuitively, we will see later that $\mathbb{N} \sim \mathbb{Z}$ and $\mathbb{Z} \sim \mathbb{Q}$.

If we have any collection (set) of sets, $\sim$ will be a relation on this collection; in fact it will be an equivalence relation:

Theorem 6.1: Let $\mathbb{A}$ be a collection of sets. Then $\sim$ is an equivalence relation on $\mathbb{A}$.

Proof: We must show that $\sim$ is reflexive, symmetric and transitive. Let $A \in \mathbb{A}$. Since $I_A : A \to A$ is a bijection, $A \sim A$ and $\sim$ is reflexive. Next, suppose that $A, B \in \mathbb{A}$ with $A \sim B$. Then there exists a one-to-one correspondence $f : A \to B$. But $f^{-1} : B \to A$ is also a one-to-one correspondence (theorem 2.14) so $B \sim A$ and $\sim$ is symmetric. To show transitivity, suppose A, B and C are elements of $\mathbb{A}$ with $A \sim B$ and $B \sim C$. Then there exist one-to-one correspondences $f : A \to B$ and $g : B \to C$. But $g \circ f : A \to C$ is also a one-to-one correspondence (theorem 2.15) so $A \sim C$ and $\sim$ is transitive and hence is an equivalence relation. $\qquad \square$

We know that an equivalence relation on a set induces a partition of that set; the equivalence classes induced by $\sim$ are just collections of sets all of which have the same number of elements (see exercise 1 below).

It will be convenient to have a notation for certain sets of natural numbers; if $k \in \mathbb{N}$ we will denote $\{1, 2, \ldots, k\}$ by N_k. For example, $N_3 = \{1, 2, 3\}$, $N_7 = \{1, 2, 3, 4, 5, 6, 7\}$. We also define $N_0 = \varnothing$. We can think of N_k as the archetypical set with k elements, for if A is any set such that $A \sim N_k$ then we say that A has k elements.

The concept we will use for "the number of elements in a set" is that of *cardinal number*. We won't define what a cardinal number is (it will remain an undefined concept) but we will list some properties (axioms) which cardinal numbers satisfy:

Axiom 1. Each set A has a cardinal number associated with it (which we will call the *cardinal number of the set*), denoted by $|A|$. For each cardinal number x there exists a set X such that $x = |X|$.

Axiom 2. $|A| = 0$ if and only if $A = \varnothing$.

Axiom 3. If $A \sim N_k$ then $|A| = k$.

Axiom 4. $|A| = |B|$ if and only if $A \sim B$.

We see that, for finite sets anyway, the cardinal number of a set is just the number of elements in the set; that is, a non-negative integer. What happens in the case of infinite sets remains to be seen, but we can think of cardinal numbers as symbols representing the number of elements in a set. Putting this yet another way, every element in an equivalence class of $\sim$ has the same cardinal number.

Next we wish to define a relation on sets of cardinal numbers.

Definition 6.2: Let A and B be sets. We say $|A| \leq |B|$ if and only if there exists an injection $f: A \to B$. If $|A| \leq |B|$ but $|A| \neq |B|$ then we say $|A| < |B|$.

For example, if $A = \{1, 3, 5, 7\}$ then $|A| < |N_5|$, $|A| \leq |N_4|$ and $|N_2| \leq |A|$.

A useful observation about $\leq$ is that if $f: A \to B$ is an injection then $A \sim Im(f)$ and as $Im(f) \subseteq B$ we could have defined "$|A| \leq |B|$" as "A is equivalent to a subset of B."

The notation for $\leq$ suggests that it is a partial order and this is indeed the case:

Theorem 6.2: Let $\mathbb{C}$ be a set of cardinal numbers. Then $\leq$ is a partial order on $\mathbb{C}$; that is, $\leq$ is reflexive, transitive and antisymmetric.

Outline of Proof: Let $x \in \mathbb{C}$ and let A be a set such that $|A| = x$. Since $I_A: A \to A$ is an injection, $|A| \leq |A|$ so $\leq$ is reflexive. Now suppose that x, y and $z \in \mathbb{C}$, with $x \leq y$ and $y \leq z$. Let A, B and C be sets such that $|A| = x$, $|B| = y$ and $|C| = z$. Then there exist f, g injections such that $f: A \to B$ and $g: B \to C$. But we know (theorem 2.16) that $g \circ f: A \to C$ is an injection, so $x \leq z$ and $\leq$ is transitive. At first glance it appears "obvious" that $\leq$ is antisymmetric, for if A is equivalent to a subset of B and B is equivalent to a subset of A then it would seem that A must be equivalent to B. This is clearly the case for finite sets, but we must remember that it must be proven for any sets, finite or infinite. Actually, it is surprisingly difficult to prove that $\leq$ is antisymmetric; in fact, this result is even graced with some names: The Schröder-Bernstein theorem (or the Cantor-Bernstein theorem) and was proven independently in the 1890's by Schröder and Bernstein (it turns out that Schröder's proof was incorrect). Cantor had proven a special case earlier and conjectured the general case. We won't give a proof here; the interested reader is referred to, for example, *Naive Set Theory* by Paul R. Halmos, Springer-Verlag, New York, 1974. $\square$

Exercises 6.2

1. Let $A = \{1, 3, 5, 7\}$ and consider $\sim$ on $\mathbb{P}(A)$.
   a) Let $B = \{1, 5\}$. Find two sets which are equivalent to B and two sets which are not equivalent to B.
   b) Find two sets in $\mathbb{P}(A)$ which are equivalent to N_3.
   c) Find $[\{1, 3, 5\}]_\sim$.
   d) Find $[\mathbb{P}(A)]_\sim$.

2. Prove the observation which was made after definition 6.2. That is, let A and B be sets with $|A| \leq |B|$. Show that A is equivalent to a subset of B.

3. Suppose that A, B, C are sets with $|A| < |B|$ and $B \sim C$. Show that $|A| < |C|$.

4. Suppose that A, B are sets with $A \subseteq B$. Show that $|A| \leq |B|$.

5. Show that $N_0 \nsim N_1$.

6. *Believe It or Not*: Conjecture: Let A, B be sets with $|A| \leq |B|$. Then $A \subseteq B$.

   "Proof": Let A, B be as above. Since $|A| \leq |B|$, there exists an injection $f : A \to B$. Let $C = f(A)$. Clearly, $C \subseteq B$. Also, if $x \in A$ then $f(x) \in C$ so $A \subseteq C$. Hence, $A \subseteq B$. $\square$

   "Counterexample": Let $A = \{1, 2, 3, 4\}$, $B = \{1, 2, 4, 5\}$. Then $|A| = |B|$ but $A \nsubseteq B$.

7. *Believe It or Not*: Conjecture: Let A, B be sets with $|A| < |B|$. Then $A \neq B$.

   "Proof": Let A, B be as above. Since $|A| < |B|$, there exists an injection $f : A \to B$. But $|A| \neq |B|$ so f is not onto B. Let $x \in B - f(A)$. Clearly $x \in B$ and $x \notin A$ so $A \neq B$. $\square$

   "Counterexample": Let $A = \{n \in \mathbb{N} : n \text{ is even}\}$, $B = \mathbb{N}$. Then $|A| < |B|$ but $A \neq B$.

6.3   INFINITE SETS

We have been using the terms *finite set* and *infinite set* without really knowing what they mean. We would probably agree that for each $k \in \mathbb{N}$, N_k is a finite set and that $\mathbb{N}$ and $\mathbb{R}$ are infinite sets, but what about a precise definition? The definition which we will use may seem strange at first, but we will see that it gives the desired results; that is, it agrees with our intuitive notion of finiteness and infiniteness.

Definition 6.3: Let A be a set. A is *infinite* if and only if there exists a proper subset of A which is equivalent to A. (In symbols, A is infinite iff $\exists B \subset A$ such that $A \sim B$.) A set which is not infinite is *finite*.

As an example of this, let $f : \mathbb{N} \to E$ be given by $f(n) = 2n$, where $E = \{n \in \mathbb{N} : n \text{ is even}\}$. Then f is a one-to-one correspondence (see exercise 1 below) so $\mathbb{N} \sim E$. Thus $\mathbb{N}$ is an infinite set, a somewhat encouraging fact.

It should be easy to see (exercise 2 below) that A is infinite if and only if there exists a proper subset B of A and an injection $f : A \to B$.

The following sequence of theorems will show that our definition of infinite and finite does in fact agree with our intuitive notion of these concepts. It may be helpful to read over the statement of all the theorems first before returning to look at the proofs so you can get an idea of the path being taken.

Theorem 6.3: Let A and B be sets with $A \subseteq B$. If A is infinite then B is infinite.

Proof: Suppose that A is an infinite set and let B be any set such that $A \subseteq B$. Since A is infinite there exists an injection $f : A \to C$ where $C \subset A$. We now define $g : B \to B$ by

$$g(x) = \begin{cases} f(x), & \text{if } x \in A; \\ x, & \text{if } x \in B - A. \end{cases}$$

The picture below may help to see what is going on here; the $Im(g)$ is shaded.

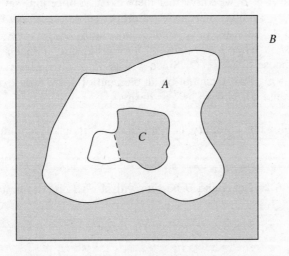

We will have shown B to be infinite if we can show that g is an injection and $Im(g) \subset B$. That g is an injection is left to exercise 3. We note that $(A - C) \cap Im(g) = \varnothing$, so to show that $Im(g) \subset B$, let $y \in A - C$ (such a y exists since $C \subset A$). Thus $y \in A$ and hence $y \in B$, so $y \notin B - A$. Hence for any $x \in A$, $y \neq f(x)$ and thus $y \notin Im(g)$ so $Im(g)$ is a proper subset of B. $\qquad \square$

An immediate corollary of this is

Corollary 6.1: Let A and B be sets with $A \subseteq B$. If B is finite then A is finite.

Proof: Since "finite" is the negation of "infinite," this is just the contrapositive of theorem 6.3. $\qquad \square$

Thus we see that subsets of finite sets are finite and supersets of infinite sets are infinite.

Next, we show that if two sets are equivalent then they must both be finite or infinite.

Theorem 6.4: Let A and B be sets with $A \sim B$. If A is infinite then B is infinite.

Proof: Let A, B be as above, with A infinite. We must find an injection from B to a proper subset of B. We will do this by using some bijections and injections which we already have (for an example of this, see exercise 4 below). Since $A \sim B$ we know that there exists a bijection, say $f: A \rightarrow B$. Also, since A is infinite, there exists a proper subset of A, say C, which is equivalent to A. Thus we have another bijection, say $g: A \rightarrow C$. Now let $h = f \circ g \circ f^{-1}$. Then $h: B \rightarrow B$ and h is an injection (since f, f^{-1} and g are). Now, choose $z \in A - C$. Since f^{-1} is onto, there exists $x \in B$ such that $f^{-1}(x) = z$. This x cannot be an element of $Im(h)$ (do exercise 4 for the proof of this) so $Im(h) \subset B$, as desired. $\qquad \square$

The following corollary is immediate as it is the contrapositive of theorem 6.4:

Corollary 6.2: Let A and B be sets with $A \sim B$. If A is finite then B is finite.

The next theorem seems "obvious" (based on our intuitive understanding of infinite sets) but its proof is surprisingly intricate:

Theorem 6.5: If A is an infinite set and $x_0 \in A$ then $A - \{x_0\}$ is infinite.

Proof: Let A be an infinite set and x_0 an element of A. To show $A - \{x_0\}$ is infinite, we must find an injection f from $A - \{x_0\}$ to a proper subset of $A - \{x_0\}$. To construct this injection, we will use the fact that A is infinite, so there exists an injection, say $g: A \rightarrow A$ with $g(A) \subset A$. We consider three cases:

Case I. $x_0 \notin g(A)$. In this case, define $f: A - \{x_0\} \rightarrow A - \{x_0\}$ by $f(x) = g(x)$. Then f is clearly an injection and $g(x_0) \neq x_0$ so $g(x_0) \notin Im(f)$ and $Im(f) \subset A - \{x_0\}$.

Case II. $x_0 \in g(A)$ and $g(x_0) = x_0$. Again we can define $f: A - \{x_0\} \rightarrow A - \{x_0\}$ by $f(x) = g(x)$. Now let $x_1 \in A - g(A)$. Then $x_1 \in A - \{x_0\}$ but $x_1 \notin Im(f)$ (details left as exercise 6), so $A - \{x_0\}$ is infinite.

Case III. $x_0 \in g(A)$ and $g(x_0) \neq x_0$. Let $x_1 = g^{-1}(x_0)$ and $x_2 = g(x_0)$. Now we define $f: A - \{x_0\} \rightarrow A - \{x_0\}$ by

$$f(x) = \begin{cases} g(x), & \text{if } x \neq x_1; \\ x_2, & \text{if } x = x_1. \end{cases}$$

The only way f can fail to be an injection is for x_2 to have another preimage in addition to x_1, but the only possible candidate is x_0, which is not in the domain of f; therefore, f is an injection. Now, let $x_3 \in A - g(A)$. Then $x_3 \in A - \{x_0\}$ but $x_3 \notin Im(f)$ (details left as exercise 7); hence, $A - \{x_0\}$ is infinite.

As our three cases are exhaustive, this completes the proof. $\square$

The next two theorems characterize finite sets.

Theorem 6.6: If k is a non-negative integer then N_k is a finite set.

Proof: $N_0 = \varnothing$ is finite for it has no proper subsets. Also N_1 is finite for it is not equivalent to its only proper subset, $\varnothing$ (this was shown in exercise 6 of the previous section). We will use mathematical induction to complete the proof. Suppose that for some $m \geq 1$ we have N_m finite. If N_{m+1} is infinite, then by theorem 6.5 $N_{m+1} - \{m + 1\} = N_m$ would also be infinite, a contradiction of our induction hypothesis. Hence N_m finite implies that N_{m+1} is finite, which completes our induction proof. $\quad\square$

Now all that remains to be done in our program is to show that any finite set is equivalent to some N_k:

Theorem 6.7: If A is a finite set then $A \sim N_k$ for some non-negative integer k.

Proof: The proof will be a contrapositive proof; we will assume that A is not equivalent to any N_k and show that this implies that A is an infinite set. If A is not equivalent to any N_k then certainly $A \neq \varnothing$ so we may choose an element of A, say a_1. Now $A - \{a_1\} \neq \varnothing$ (for if it were then $A \sim N_1$, a contradiction of our hypothesis) so we may continue the process and choose an element of it, say a_2. $A - \{a_1, a_2\} \neq \varnothing$ so we can do it again; in fact for any $k \in \mathbb{N}$ we can find a subset $\{a_1, a_2, \ldots, a_k\}$ of A with k elements such that $A - \{a_1, a_2, \ldots, a_k\} \neq \varnothing$. Let $B = \{a_i : i \in \mathbb{N}\}$, the set of all such elements in A. $B \sim \mathbb{N}$ (the proof of this is left as exercise 9) so B is infinite. But $B \subseteq A$, so by theorem 6.3 A is infinite. $\quad\square$

We have now completed our program of showing that our definition of finite sets and infinite sets agrees with our intuitive notions; a set is finite if and only if it has a non-negative integer for a cardinal number. We could have used theorem 6.7 as our definition of a finite set and proceeded in the opposite direction, ending with a proof of definition 6.3 that a set is infinite if and only if it is equivalent to a proper subset of itself. [The interested reader might want to carry out this approach.]

If A is an infinite set we will call $|A|$ an *infinite cardinal number*. We will use the (standard) symbol $\aleph_0$ (pronounced "aleph null"—$\aleph$ is the first letter in the Hebrew alphabet) as the cardinal number of $\mathbb{N}$. Clearly, for all finite cardinal numbers k, we have $k < \aleph_0$. Are there any other infinite cardinal numbers? The answer to this question will have to wait until the next section.

Exercises 6.3

1. Show that $f: \mathbb{N} \to E$ given by $f(n) = 2n$ is a one-to-one correspondence ($E = \{n \in \mathbb{N} : n \text{ is even}\}$).

2. Show that A is an infinite set if and only if there exists a proper subset B of A and an injection $f : A \rightarrow B$.

3. Show that the function g in the proof of theorem 6.3 is an injection.

4. Show that E (the set of even natural numbers) is infinite by explicitly constructing the injection h as in the proof of theorem 6.4 by using f (from exercise 1) twice.

5. Complete the proof of theorem 6.4 by showing that $x \notin Im(h)$.

6. Complete the proof of case II in theorem 6.5 by showing that x_1 has the desired properties.

7. Complete the proof of case III in theorem 6.5 by showing that x_3 has the desired properties.

8. Show that if A is an infinite set and B is a finite set then $A - B$ is also infinite.

9. Show that the set B in the proof of theorem 6.7 is indeed equivalent to $\mathbb{N}$ by finding a bijection from B to $\mathbb{N}$.

10. Show that if k is a finite cardinal number then $k < \aleph_0$.

11. Show that if A and B are finite sets then $A \cup B$ and $A \cap B$ are finite.

12. ***Believe It or Not:*** Conjecture: If $A \sim N_r$ and $B \sim N_s$ then $A \cup B \sim N_{r+s}$.

   "Proof": Let $A \sim N_r$ and $B \sim N_s$. Then there exist bijections f, g such that $f : A \rightarrow N_r$ and $g : B \rightarrow N_s$. Consider $h : A \cup B \rightarrow N_{r+s}$ defined by

$$h(x) = \begin{cases} f(x), & \text{if } x \in A; \\ g(x) + r, & \text{if } x \in B. \end{cases}$$

   We see that h is onto N_{r+s} for if $k \in N_{r+s}$ and $k \leq r$ then there exists $x \in A$ such that $f(x) = h(x) = k$. If $r < k \leq r + s$ then $k - r \in N_s$ so there exists $x \in B$ such that $g(x) = k - r$ so $h(x) = k$. Now suppose that $x, y \in A \cup B$ with $h(x) = h(y)$. If $h(x) \in N_r$ then $h(x) = f(x) = f(y)$ so $x = y$ since f is an injection. Similarly, if $r < h(x) \leq r + s$ then $h(x) = g(x) + r = g(y) + r$ so $x = y$ and h is an injection. $\qquad \square$

   "Counterexample": Let $A = \emptyset, B = \{a, b, c\}$. Clearly, $A \cup B \sim N_3$, not $N_{1+3} = N_4$.

13. ***Believe It or Not***: Conjecture: Let A, B be sets with $A \cup B$ infinite. Then at least one of A, B is infinite.

   "Proof": Suppose that $A \cup B$ is infinite. Then there exists a bijection $f : A \cup B \rightarrow C$ where $C \subset A \cup B$. Consider $g = f|_A$. Then g is clearly a bijection and since $g(A) \subset C$, $g(A) \subset A$ so A is infinite. $\qquad \square$

"Counterexample": Let $A \cup B = \mathbb{N}$ with $A = \{n \in \mathbb{N} : n \leq \mathbb{N}/2\}$, $B = \{n \in \mathbb{N} : n \geq \mathbb{N}/2\}$. Then A, B are both finite but $A \cup B$ is infinite.

6.4 INFINITE CARDINAL NUMBERS

Now that we know what finite and infinite sets are and we have a concept for the "number of elements in a set" (cardinal numbers) we can consider the question "Do all infinite sets have the same number of elements?" The surprising answer to this question is "No!" We will first see that there are an infinite number of distinct infinite cardinal numbers and then determine the cardinal numbers of some familiar sets. Our intuition will probably not be a good guide in this endeavor since it is based on our experience with finite sets and as we have seen, infinite sets behave somewhat differently (infinite sets are equivalent to proper subsets, for example).

One property which they do share though is the following: it is easy to see that finite sets have fewer elements than their power sets (the one-element subsets are in one-to-one correspondence with the elements of the set and there are other subsets besides). It turns out that there are "enough" extra subsets to make this property extend to infinite sets as we see in

---

Theorem 6.8: Let A be a set. Then $|A| < |\mathbb{P}(A)|$.

Proof: If $A = \varnothing$ then $\mathbb{P}(A) = \{\varnothing\}$ so $|A| = 0$ and $|\mathbb{P}(A)| = 1$; hence, $|A| < |\mathbb{P}(A)|$. Next, suppose that $A \neq \varnothing$. We define $f : A \to \mathbb{P}(A)$ by $f(x) = \{x\}$. Clearly f is an injection so we have $|A| \leq |\mathbb{P}(A)|$, and all we need to show is that $|A| \neq |\mathbb{P}(A)|$, which is the same as $A \nsim \mathbb{P}(A)$. Suppose, to the contrary, that $A \sim \mathbb{P}(A)$. Thus there exists a one-to-one correspondence $g : A \to \mathbb{P}(A)$. Let $S = \{x \in A : x \notin g(x)\}$. Then $S \in \mathbb{P}(A)$ and as g is onto, there exists $y \in A$ such that $g(y) = S$. Is $y \in S$? If so, then $y \notin g(y)$ which means that $y \notin S$, an impossibility. Thus we must have $y \notin S$. But this means that $y \in g(y)$ which, because of the way in which S was defined, implies $y \in S$, a contradiction. Thus g cannot be onto and thus we have $|A| < |\mathbb{P}(A)|$. $\qquad \square$

This theorem gives us an infinite sequence of distinct infinite cardinal numbers:

$$|\mathbb{N}| < |\mathbb{P}(\mathbb{N})| < |\mathbb{P}(\mathbb{P}(\mathbb{N}))| < \cdots.$$

It will be helpful to give a name to sets which have no more elements than $\mathbb{N}$:

> **Definition 6.4:** If set A is equivalent to $\mathbb{N}$ we say that A is *denumerable*. If A is denumerable or finite we say that A is *countable*. If A is infinite but not denumerable, we say that A is *uncountable*.

Thus sets that can be put into one-to-one correspondence with subsets of $\mathbb{N}$ are countable. Some other examples of countable sets which may or may not be so obvious are

1. The set of even natural numbers.

2. The set of odd natural numbers.

3. $\mathbb{Z}$, the set of integers.

4. $\mathbb{N} \times \mathbb{N}$.

5. $\mathbb{Q}$, the set of rational numbers.

6. The set of algebraic numbers, i.e., the set

$$\{x : x \text{ is a root of a polynomial with integer coefficients}\}.$$

The verification of some of these statements is left to the exercises.

We might suspect (as Cantor did for awhile) that most of the infinite sets with which we are familiar are denumerable, but this is not the case, as we see in

> **Theorem 6.9:** The open unit interval, $(0, 1) = \{x : x \in \mathbb{R}, 0 < x < 1\}$, is uncountable.

Proof: $(0, 1)$ is clearly infinite so we will suppose that it is denumerable; that is, there is a one-to-one correspondence, say $f : \mathbb{N} \to (0, 1)$. We will use the fact that every element in $(0, 1)$ can be written as a decimal

$$0.x_1 x_2 x_3 \cdots \text{ where each } x_i \in \{0, 1, \ldots, 9\}.$$

If for a particular number the decimal expansion terminates in 0's we will agree to decrease the last non-zero entry by 1 and replace the 0's by 9's. Thus $0.25000 \cdots$ will be written as $0.24999 \cdots$. With this stipulation two numbers will be equal if and only if their decimal expansions agree in every place (you might try to prove this assertion). (Note: The argument which we are about to use has come to be known as "Cantor's diagonalization argument," but interestingly enough it was not used in his first proof of this theorem.) Since we have assumed that $(0, 1) \sim \mathbb{N}$ we can list the elements as, say, $a_1, a_2, \ldots$ (here we can think of $a_1 = f(1)$, $a_2 = f(2)$, $\ldots$). Suppose that we have listed all the elements of $(0,1)$ in this way:

$$a_1 = 0.x_{11}x_{12}x_{13}\cdots$$
$$a_2 = 0.x_{21}x_{22}x_{23}\cdots$$
$$a_3 = 0.x_{31}x_{32}x_{33}\cdots$$

.

.

$$a_n = 0.x_{n1}x_{n2}x_{n3}\cdots$$

.

.

Now consider the number

$$a = 0.x_1x_2x_3\cdots$$

where

$$x_i = \begin{cases} 2, & \text{if } x_{ii} \neq 2; \\ 1, & \text{if } x_{ii} = 2. \end{cases}$$

Then clearly $a \in (0, 1)$ but it does not appear on our list since it differs from each a_i in the ith place. Therefore our list (actually the image of f) is not complete; i.e., f is not onto. Therefore, $(0, 1)$ is uncountable. □

In terms of cardinal numbers, this means that

$$\aleph_0 < |(0, 1)|.$$

Since $(0, 1) \subseteq \mathbb{R}$ we see that $\mathbb{R}$ is uncountable (see exercise 10 below for the details), but is $|(0, 1)| < |\mathbb{R}|$? The answer is no:

Theorem 6.10: $(0, 1) \sim \mathbb{R}$.

Proof: The function $f: (0, 1) \to \mathbb{R}$ given by $f(x) = \tan \pi(x - \frac{1}{2})$ is a bijection (details left as exercise 11) so $(0, 1) \sim \mathbb{R}$. □

Since $\mathbb{R}$ is uncountable its cardinal number is not $\aleph_0$; its cardinal number is usually denoted by c (for continuum.) Are there any cardinal numbers, say b, such that

$$\aleph_0 < b < c \; ?$$

This question has received much attention from mathematicians since the turn of the century. It has been shown that using the usual axioms of set theory the existence of such cardinal numbers can neither be proven or disproven; that is, one can assume that such cardinals exist or assume that they do not. The assumption that no such cardinals exist is called the continuum hypothesis. What has been shown is that $|\mathbb{P}(\mathbb{N})| = c$. This gives rise to a more general question; if A is an infinite set, does there exist a cardinal number, say b, such that

$$|A| < b < |\mathbb{P}(A)| ?$$

The existence of such cardinal numbers has also been shown to be independent of the usual axioms of set theory and the assumption that no such cardinals exist is called the *generalized continuum hypothesis*. [Clearly such cardinal numbers exist if A is a non-empty finite set.]

Exercises 6.4

1. Show that the f defined in the proof of theorem 6.8 is indeed an injection.

2. Show that the set of odd natural numbers is denumerable.

3. Show that the set of integers is denumerable.

4. Show that the set of all integer multiples of 10^6 is denumerable.

5. Show that $\mathbb{Q}$ is denumerable, assuming that $\mathbb{N} \times \mathbb{N}$ is denumerable.

6. Show that any subset of a countable set is countable.

7. Give three examples of algebraic numbers which are not rational numbers.

8. Show that the decimal representation of numbers (with the stipulation mentioned in the proof of theorem 6.9) is unique.

9. Suppose that the first five numbers on our list of elements of $(0,1)$ (as in the proof of theorem 6.9) were

$$a_1 = 0.12115 \cdots$$

$$a_2 = 0.34221 \cdots$$

$$a_3 = 0.99415 \cdots$$

$$a_4 = 0.55789 \cdots$$

$$a_5 = 0.22391 \cdots.$$

Write the first five places in the decimal representation of the number a.

10. Show that if $A \subseteq B$ and A is uncountable, then B is uncountable.

11. Show that f given in the proof of theorem 6.10 is a bijection.

12. ***Believe It or Not***: Conjecture: Suppose that A and B are countable sets. Then $A \cup B$ is countable.

"Proof": If both A and B are finite, then by exercise 11 in the previous section $A \cup B$ is finite and hence countable. Suppose that both A and B are denumerable. Then we can list their elements: $A = \{a_1, a_2, \ldots\}, B = \{b_1, b_2, \ldots\}$. Thus $A \cup B = \{a_1, b_1, a_2, b_2, \ldots\}$, which is clearly denumerable. □

"Counterexample": Let $A = \mathbb{Q}$ and $B = \mathbb{N} \times \mathbb{N}$. Then $A \cup B$ is in one-to-one correspondence with $\mathbb{N} \times \mathbb{N} \times \mathbb{N} \times \mathbb{N}$ which is uncountable.

ANSWERS AND HINTS FOR SELECTED EXERCISES

1.2 1. c) F ("but" has same meaning in logic as "and").

d) T.

2. a) ii) $\neg(\neg p \vee q)$ F.

4. a) $3 - 4 \geq 7$.

5. b)

p	q	$q \star p$
T	T	F
T	F	T
F	T	F
F	F	F

1.3 1. Partial answer: a), c) and f) are logically equivalent.

3. b) Use truth tables or observe that when p is T and q is F then $\neg(p \vee q)$ is F while $\neg p \vee \neg q$ is T.

4. b) $p \rightarrow \neg q$.

d) $p \wedge q$.

6. a) If $2 = 3$ then $4 < 2$.

c) No such example exists as an implication with a true conclusion must be true.

d) If $2 + 2 = 4$ then $2 < 1$.

9. c) True.

f) True.

1.4   2. b) This is not a tautology so it is not on the list.
        e) 13 (contrapositive).
     3. a) If $2 < 1$ and $3 = 2 + 1$ then $2 < 1$.
     4. d) Can disprove using a truth table or by observing that when p is
        F and q is T then $(p \rightarrow q) \wedge \neg p$ is T but $\neg q$ is F.
     5. b) **c.**
        d) **t.**
     9. d) Is a tautology.
        e) Is not a tautology.

1.5   1. b) Invalid: can show using a truth table or by observing that when
        p is T, q is T and r is T the hypotheses are all true but the
        conclusion is F.
     2. e) No such example exists because if an argument is valid and the
        hypotheses are true then the conclusion must also be true.
        g) $2 + 2 = 5$
                    $\underline{2 + 2 = 5 \text{ implies } 1 < 3}$
                       $1 < 3.$

     3. c) Invalid, for if p is T, r is T and q is T then all the hypotheses
        are T but the conclusion is F.
        e)

1. $\neg(p \rightarrow q)$	Hypothesis (indirect proof)
2. $p \wedge \neg q$	Logical consequence (L.C.) of 1
3. p	L. C. of 2
4. $\neg p$	Hypothesis
5. $p \wedge \neg p$	L. C. of 3 and 4
6. $p \rightarrow q$	L. C. 5 (indirect proof)

        k)

1. $\neg r$	Hypothesis
2. $p \rightarrow r$	Hypothesis
3. $\neg r \rightarrow \neg p$	L. C. of 2
4. $\neg p$	L. C. of 1 and 3
5. $p \vee q$	Hypothesis
6. q	L. C. of 4 and 5

        k) (indirect)

1. $\neg q$	Hypothesis (indirect proof)
2. $p \vee q$	Hypothesis
3. p	L. C. of 1 and 2
4. $\neg r$	Hypothesis
5. $p \rightarrow r$	Hypothesis
6. $\neg r \rightarrow \neg p$	L. C. of 5
7. $\neg p$	L. C. of 4 and 6
8. $p \vee \neg p$	L. C. of 3 and 7
9. q	L. C. of 8 (indirect proof)

1.6 1. d) Let D be the set of all students and $p(x)$ be "x likes logic."
Then the proposition is $\forall x$ in $D, p(x)$.

   k) Let D be the set of integers, $p(x)$ be "x is even" and $q(x)$
be "x is divisible by 3." Then the proposition is $\exists x$ in $D \ni$
$p(x) \lor q(x)$.

   p) Let D be the set of real numbers, $p(x)$ be "x is a solution of
$x^2 - 4 = 0$," $q(x)$ be "x is positive." Then the proposition is
$\forall x$ in $D, p(x) \rightarrow q(x)$.

2. d) $\exists x$ in $D \ni \neg p(x)$. There is a student who dislikes logic.

   k) $\forall x$ in $D, \neg(p(x) \lor q(x))$. All integers are odd and not divisible
by 3.

   p) $\exists 3x$ in $D \ni \neg(p(x) \rightarrow q(x))$. There is a solution of $x^2 - 4 = 0$
which is not positive.

3. c) Every even natural number is divisible by 3; false; there is an
even natural number which is not divisible by 3.

   i) There is a natural number such that if it is divisible by 3 then
the next natural number is also divisible by 3; true; all natural
numbers are divisible by 3 and the next natural number is not
divisible by 3.

4. c) If D' is the set of natural numbers divisible by 3 then this proposition will be true.

5. b) Sometimes correct; it will be true when D contains at most one
element or if $p(x)$ is true for all x in D and it will be false if
there is at least one element x in D for which $p(x)$ is true and
one element in D for which it is false.

1.7 1. f) Let D be the set of integers, $p(x)$ be "x is odd." Then the
proposition is $(\forall x$ in $D, p(x)) \rightarrow \forall x$ in $D, \neg p(x)$.

2. f) Every integer is odd and there exists an odd integer.

3. c) For every natural number x and every natural number y, $x + 2 > y$. This is false.

4. c) [1 is even or 1 is odd] and [2 is even or 2 is odd].

5. b) Let D be the set of natural numbers, $p(x)$ be "x is even" and
$q(x)$ be "x is odd."

1.8 1. b) Direct: Let f be a differentiable function.

                             $\vdots$

                   Thus f is continuous.

          Contrapositive: Suppose that f is not a continuous function.

                             $\vdots$

                   Thus f is not differentiable.

Indirect: Suppose that f is a differentiable function which is not continuous.

$$\vdots$$

Some contradiction.

e) Direct: Let H be the homomorphic image of the cyclic group G.

$$\vdots$$

Thus H is cyclic.

Contrapositive: Suppose that H is not a cyclic group.

$$\vdots$$

Thus H is not the homomorphic image of any cyclic group.

Indirect: Suppose that H is the homomorphic image of a cyclic group and H is not cyclic.

$$\vdots$$

Some contradiction.

2. c) Incorrect. $x - y$ odd is the negation of the conclusion so to use the contrapositive method of proof we would have to show that x is odd or y is odd, which this "proof" does not do.

e) Correct indirect proof.

3. a) (Partial answer) Direct: Suppose that x is even and y is odd. Then there exist integers j, k such that $x = 2k$ and $y = 2j + 1$. Thus

$$x + y = 2k + 2j + 1 = 2(k + j) + 1$$

and hence $x + y$ is odd.

Contrapositive: Suppose $x + y$ is even. Then there exists an integer k such that $x + y = 2k$. If x is odd we are done so suppose that x is even, say $x = 2j$ for some integer j. Then $y = 2k - 2j = 2(k - j)$ so y is even and we are done.

4. a) False, for example, let $x = 3$.

c) True. Suppose x is odd, say $x = 2k + 1$ for some integer k. Then

$$x^2 = (2k + 1)^2 = 4k^2 + 4k + 1 = 2(2k^2 + 2k) + 1$$

is odd.

[Note: We use the contrapositive proof here, as a direct proof runs into trouble when we try to go from x^2 to x.]

2.1 1. c) $C - A = \{5, 7\}$.
      f) $A^C \cap C^C = \{6, 8\}$.
   3. b) $[-1, 0]$.
      d) $[0, 11)$.
      l) $\{\emptyset, \{1\}\}$.
      m) This answer requires lots of paper!
   5. a) First we show that $A \cup \emptyset \subseteq A$. Let $x \in A \cup \emptyset$. Then $x \in A$ or $x \in \emptyset$. But $x \notin \emptyset$ so $x \in A$, thus $A \cup \emptyset \subseteq A$. Now suppose that $x \in A$. Then $x \in A$ or $x \in \emptyset$ so $x \in A \cup \emptyset$. Therefore, $A \cup \emptyset = A$.
      p) Let $x \in A \cup (B - A)$. If $x \in A$ we are done so suppose that $x \in B - A$. Then $x \in B$ and $x \notin A$. Thus $x \in A \cup B$ and we have $A \cup (A - B) \subseteq A \cup B$. Now suppose that $x \in A \cup B$. Then $x \in A$ or $x \in B$. If $x \in A$ we are done so suppose that $x \notin A$. Then $x \in B$ and hence we have $x \in B - A$ and the proof is complete.
   6. c) First we show that $A \subseteq B \rightarrow A \cup B = B$. Let $x \in B$. Then $x \in A$ or $x \in B$ so $x \in A \cup B$. Now let $x \in A \cup B$. Hence $x \in A$ or $x \in B$. If $x \in B$ we are done so suppose that $x \in A$. Since $A \subseteq B$ we have $x \in B$ and thus $A \cup B = B$. Now we show that $A \cup B = B \rightarrow A \subseteq B$. Let $x \in A$. Then $x \in A \cup B$ and therefore, $x \in B$ as $A \cup B = B$.
   7.   Hint—The conjecture is true.
   8.   Hint—The conjecture is false.
  10. c) $A \cap (A^C \cup B) = (A \cap A^C) \cup (A \cap B) = \emptyset \cup (A \cap B) = A \cap B$.
  12. c) Hint—This is false.
2.2 1. b) $\{1, 3, 4, 5, 6, 7, 8\}$.
      f) True.
   3. a) The first will be true when $P \cup Q = D$; the second will be true when $P = D$ or $Q = D$. Certainly the second condition implies the first. If we let $D = \mathbb{N}$, $p(x)$ be "x is even," and $q(x)$ be "x is odd" then the first is true but the second is false.
2.3 2. a) $R = \{(1, 2), (1, 3), (2, 3)\}$. R is transitive, antisymmetric, ir-reflexive, complete and asymmetric.
   3. d) True. Let $(x, y) \in A \times C$. Then $x \in A$ and $y \in C$. But since $A \subseteq B$ and $C \subseteq D$ we have $x \in B$ and $y \in D$. Thus $(x, y) \in B \times D$.
   4. b) Let $x \in A$. ... Then $(x, x) \notin R$.
   6. a) (Partial answer) Since $R = \{(x, y) : x, y \in A\}$, certainly $\{(x, x) : x \in A\} \subseteq R$. If A contains more than one element, R won't be asymmetric or antisymmetric.
   7. a) R is not reflexive since no one is his (or her) own parent. If xRy then y is x's parent which means that x is not y's parent so $\neg(xRy)$ is true and R is asymmetric.

8. a) $\{1, 6, 11, 16, \ldots\}$.
9. b) (Partial answer) $(0, 1) \notin R$.
10. a) False. Let $A = \{1, 2\}$ and $R = \{(1, 2), (2, 1)\}$.
    h) True. Let $(x, y) \in R \cap S$. Then (x, y) is in both R and S. Since they are both symmetric, we have (y, x) in both R and S so $(y, x) \in R \cap S$.
11. a) Let $A = \{1, 2, 3\}$, $R = \{(1, 1), (2, 2), (3, 3), (2, 3), (3, 2), (1, 2), (2, 1)\}$. Then $(1, 2) \in R$ and $(2, 3) \in R$ but $(1, 3) \notin R$.
12. c) Hint—This is false.
17. a) $R_{sym} = \{(1, 2), (1, 4), (2, 3), (2, 1), (4, 1), (3, 2)\}$.
19. a) $<2>_R = \{1, 4\}$.
20. Hint—The conjecture is false and so is the counterexample.

2.4 1. b) $Dom(S) = \{1, 3\}$.
    d) $Im(R) = \{3, 4\}$.
    n) $S^{-1} = \{(1, 1), (4, 3), (2, 3)\}$.
    p) $I_B = \{(1, 1), (3, 3), (4, 4)\}$.
  2. d) Suppose that R is symmetric. Let $(x, y) \in R$. Then $(y, x) \in R$ so $(x, y) \in R^{-1}$ and $R \subseteq R^{-1}$. Now suppose that $(x, y) \in R^{-1}$. Then $(y, x) \in R$ so $(x, y) \in R$ and $R^{-1} \subseteq R$ and we have $R = R^{-1}$. Next suppose that $R = R^{-1}$. Let $(x, y) \in R$. Then $(x, y) \in R^{-1}$ (since $R = R^{-1}$) so $(y, x) \in R$ and R is symmetric.
  4. b) Let $y \in Im(S \circ R)$. Then there exists an x such that $(x, y) \in S \circ R$. But this means that there exists a z such that $(x, z) \in R$ and $(z, y) \in S$. Hence $y \in Im(S)$.
  9. a) Hint—This is true.
  12. Hint—The proof is incorrect.

2.5 1. $[2]_{A/\Pi} = \{2, 4, 6\}$.
  3. a) i) $\Pi_1 = \{\{1\}, \{2, 3\}, \{4\}\}$, $\Pi_2 = \{\{1, 2, 3\}, \{4\}\}$.
    ii) $\Pi_1 = \{\{1, 2\}, \{3, 4\}\}$, $\Pi_2 = \{\{1, 3\}, \{2, 4\}\}$.
  4. (Partial answer) Let $(x, y) \in A/[A]_R$. Then x and y are both elements of the same equivalence class of R. But this means that xRy or $(x, y) \in R$. Hence $A/[A]_R \subseteq R$.
  6. a) i) Let $R_1 = \{(1, 1), (2, 2), (3, 3), (4, 4)\}$.
    $R_2 = \{(1, 1), (2, 2), (3, 3), (4, 4), (1, 2), (2, 1)\}$.
  7. a) $\Psi \star \Pi = \{\{1, 2\}, \{3\}, \{4\}, \{5\}\}$.
  8. a) $[2]_3 = \{2, 5, 8, \ldots\}$.
    b) ii) One solution is $x = 3$.
  10. Hint—Compute a few elements of $Z_2 \star Z_3$.

2.6 1. a) $f(3) = 4$.
    c) (Partial answer) If $x = 1$ then $f(6) = x$, if $x \neq 1$ then $f(x - 1) = x$ so f is onto.
  4. a) $f^{-1} \circ f = \{(1, 1), (2, 2), (3, 3), (4, 4), (2, 3), (3, 2)\}$.
  5. a) Hint—Define g so that $g(y) = x$ when $y \in Im(f)$ and where $f(x) = y$. For $y \notin Im(f)$ define $g(y)$ arbitrarily.

8. Hint—Think about what functions are reflexive.
9. b) Hint—This is not true.
13. c) Hint—This is true.
15. Hint—Proof is not correct.

2.7 1. a) $f(A - \{2\}) = \{3, 4\}$.
 2. e) (Partial answer) Let $C = \{1\}$. Then $f^{-1}(f(C)) = \{-1, 1\}$.
 3. c) (Partial answer) Suppose f is one-to-one. Let $C, D \in \mathbb{P}(A)$ with $f^*(C) = f^*(D)$. Suppose $C \neq D$. Then (say) there exists $x \in C$ such that $x \notin D$. But $f(x) \in f^*(C) = f^*(D)$ so there must be a $y \neq x$ in D such that $f(y) = f(x)$. But this contradicts the assumption that f is one-to-one.
 4. e) I_A is idempotent.
 5. c) Hint—Let x^*, y^* be inverses for x, y and compute $y^* \bullet x^* \bullet x \bullet y$.
 6. b) (Partial answer) If $A = \{1, 2\}$ then $\{1\} - \{2\} \neq \{2\} - \{1\}$.
    c) Since $B \cup B = B \cap B = B$ for all sets B, all sets are idempotent for both $\cap$ and $\cup$.
 7. Hint—It may be helpful to note that $X \bullet Y = (X \cap Y^C) \cup (Y \cap X^C)$.
 8. a) Hint—You must show that if $f, g \in F$ then $f \circ g \in F$.
    c) Hint—Think about the graphs of these functions.
13. Hint—Think about what $+$ is.
16. b) (Partial answer) $(1, 2), (2, 4), (5, 10)$ are all in $[(1, 2)]_R$.
    c) (Partial answer) Let $n \in \mathbb{Z}$, $n \neq 0$ and suppose that $(w, z) \in [(nx, ny)]_R$. Then $wny = znx$. Since $n \neq 0$ we can divide it out, obtaining $wy = zx$ so $(w, z)R(x, y)$ or $[(nx, ny)]_R \subseteq [(x, y)]_R$.

3.2 1. d) When $n = 1$ we have $1 + \frac{1}{2} \leq 2$ which is true. Now suppose that $k \in \mathbb{N}$ and $1 + 2^{-1} + \cdots + 2^{-k} \leq 2$. Multiplying this inequality by $\frac{1}{2}$ we obtain $2^{-1} + 2^{-2} + \cdots + 2^{-(k+1)} \leq 1$ and adding 1 to each side we see that the inequality holds for $k + 1$.
    h) Hint—$8 \cdot 10^{2(n+1)} + 6 \cdot 10^{2(n+1)-1} + 9 = 100(8 \cdot 10^{2n} + 6 \cdot 10^{2n-1} + 9) - 891$.
 4. Hint—Compute the first few values of a_n and try to prove a stronger result from which the desired result will follow.
 6. Hint—The proof is correct.

3.3 2. Hint—Consider the set $S = \{n \in \mathbb{Z} : n < 0\}$.
 7. Hint—Note that $\alpha\beta = -1$ and $\alpha + \beta = 1$.
 8. Hint—Use the recurrence for F_n from problem 7.

4.2 1. Suppose $a, b, c \in \mathbb{R}$ with $a \neq 0$ and $ab = ac$. Since $a \neq 0$, $a^{-1} \in \mathbb{R}$. Thus $a^{-1}(ab) = a^{-1}(ac)$ so $b = c$.
 7. Suppose that s is an upper bound for A. Let $x \in -A$. Then $-x \in A$ so $-x \leq s$. But this implies $x \geq -s$ so $-s$ is a lower bound for $-A$.

12. c) If $x \in \mathbb{R}$ with $x = y$ then $|x - y| = 0 < \epsilon, \forall \epsilon > 0$. For the converse, assume that $x \neq y$. Thus $|x - y| > 0$. Let $\epsilon = |x - y|/2$. Thus we have $|x - y| < \epsilon = |x - y|/2$, a contradiction.

18. Hint—All three are incorrect.

4.3 4. Hint—The only convergent sequences are those which are eventually constant.

8. Find $M \in \mathbb{N} \ni n \geq M$ implies $|a_n - L| < \epsilon/2$ and $|a_n - b_n| < \epsilon/2$.

11. b) (Partial answer) (a_n) defined by $a_n = n$ is not a Cauchy sequence.

d) Hint—Find $M \in \mathbb{N} \ni m, n \geq M$ implies $|a_m - a_n| < 1$. Then show $\sup(\{a_1, a_2, \ldots, a_m + 1\})$ is an upper bound for (a_n).

4.4 5. Hint—Choose $x = 1/n, z = 1/(2n)$.

12. Hint—Consider various sets for D.

17. Hint—The conjecture is true.

18. Hint—The conjecture is false.

5.2 2. Hint—Is G closed with respect to $+$?

6. Hint—Look for a large example.

8. b) Hint—Suppose $H \not\subseteq K$ and $K \not\subseteq H$ and look at $h \in H \cap K^C$ and $k \in K \cap H^C$.

16. Hint—For $b \in G$ look at $b^{-1}ab$.

21. Hint—The conjecture is false.

5.3 4. Hint—Note that in $\mathbb{Z}(4)$, $3 + 3 = 2$.

7. Hint—Not very many.

8. Hint—If $f(x) = x^2$ is an automorphism then $\forall a, b \in G$, $f(ab) = (ab)^2 = f(a)f(b) = a^2b^2$.

14. Hint—Assume G is non-abelian.

15. Hint—If $f(x) = f(y)$ then $x^{-1}y \in \ker f$.

16. Hint—Consider $f: \mathbb{R} \to \mathbb{R}_+$ defined by $f(x) = e^x$.

19. Hint—Apply theorem 5.7.

22. Hint—The conjecture is true.

6.2 4. Let A, B be sets with $A \subseteq B$. Define $f: A \to B$ by $f(x) = x$, $\forall x \in A$. Then f is clearly an injection so $|A| \leq |B|$.

7. Hint—The conjecture is true but the proof is incorrect.

6.3 12. Hint—All three are incorrect.

6.4 7. $\sqrt{2}$ is one.

12. Hint—The conjecture is true but the proof is incorrect.

INDEX